Monitoring the Environment
The Linacre Lectures 1990–91

Edited by

BRYAN CARTLEDGE
Principal of Linacre College
University of Oxford

Oxford New York Tokyo
OXFORD UNIVERSITY PRESS
1992

Oxford University Press, Walton Street, Oxford OX2 6DP
Oxford New York Toronto
Delhi Bombay Calcutta Madras Karachi
Petaling Jaya Singapore Hong Kong Tokyo
Nairobi Dar es Salaam Cape Town
Melbourne Auckland
and associated companies in
Berlin Ibadan

Oxford is a trade mark of Oxford University Press

Published in the United States
by Oxford University Press, New York

© *Oxford University Press, 1992*

All rights reserved. No part of this publication may be reproduced,
stored in a retrieval system, or transmitted, in any form or by any means,
electronic, mechanical, photocopying, recording, or otherwise, without
the prior permission of Oxford University Press

This book is sold subject to the condition that it shall not, by way
of trade or otherwise, be lent, re-sold, hired out, or otherwise circulated
without the publisher's prior consent in any form of binding or cover
other than that in which it is published and without a similar condition
including this condition being imposed on the subsequent purchaser

A catalogue record for this book is available from the British Library

Library of Congress Cataloging in Publication Data
Monitoring the environment / edited by Bryan Cartledge.
(The Linacre lectures ; 1990–91)
Includes bibliographical references and index.
1. Environmental monitoring. 2. Pollution. 3. Human ecology.
4. Ecology. 5. Nature conservation. 6. Environmental protection.
I. Cartledge, Bryan, Sir. II. Series: Linacre lecture ; 1990–91.
TD193.M68 1992 363.7–dc20 91–37355
ISBN 0–19–858408–3 (hb)
ISBN 0–19–858412–1 (pbk)

Typeset by Footnote Graphics, Warminster, Wilts.
Printed in Great Britain by
Biddles Ltd.
Guildford & King's Lynn

Acknowledgements

The Linacre Lectures could not have taken place without the financial support and practical encouragement of the Racal Electronics Group and I must express the gratitude of the College, and indeed of the University, for their generosity. ICI and the National Westminster Bank have also made valued contributions. Racal have made possible another dimension of Linacre's environmental initiative, the establishment of research studentships for three Soviet graduate scientists to undertake research in Oxford, for a year each, in environment-related fields; having experienced the environmental problems of the Soviet Union at first hand over many years, this is a project to which I attach particular importance. Racal have proved excellent partners.

I am also very grateful to Frances Morphy for taking on the formidable task of putting the lecturers' manuscripts into a state acceptable to the Oxford University Press; to Linacre students Neil Ferguson and Veronica Strang for the time and effort that they put into tracking down bibliographical details for incomplete references and providing amplification of footnotes; and to our College Secretary, Jane Edwards, for her hard work on virtually every aspect of arranging this first series of Linacre Lectures.

Oxford　　　　　　　　　　　　　　　　　　　　　　　　　　　B.G.C.
August 1991

Contents

	List of authors	viii
	Introduction *Bryan Cartledge*	1
1.	The environment: problems and prospects *Richard Southwood*	5
2.	The environment: a political view *Michael Heseltine*	42
3.	The greenhouse effect and global warming *John Mason*	55
4.	Implications of global climatic change *Crispin Tickell*	93
5.	The earth is not fragile *James E. Lovelock*	105
6.	Monitoring the ocean *John Woods*	123
7.	The dilemma of the Amazon rainforests: biological reserve or exploitable resource? *Ghillean T. Prance*	157
8.	The natural world: a global casino *John Phillipson*	193
	Index	207

Authors

Sir Bryan Cartledge
Principal, Linacre College, Oxford

Professor Sir Richard Southwood, FRS
Vice Chancellor of Oxford University and Linacre Professor of Zoology; and Chairman, National Radiological Protection Board

The Rt. Hon. Michael Heseltine, M.P.
Secretary of State for the Environment and Conservative Member of Parliament for Henley

Sir John Mason, CB, FRS
President of the University of Manchester Institute of Science and Technology

Sir Crispin Tickell
Warden of Green College, Oxford and President, The Royal Geographical Society

Professor James E. Lovelock, FRS
Author of Gaia *and* The ages of Gaia

Dr John Woods, CBE
Director of Marine and Atmospheric Sciences, Natural Environment Research Council

Professor Ghillean T. Prance
Director, Royal Botanic Gardens, Kew

Dr John Phillipson
Chairman, Royal Society for Nature Conservation

Introduction
Bryan Cartledge

If he could have understood them, Thomas Linacre would have been well satisfied with the first series of lectures on the environment which bear his name. As he lived and died long before the basic laws of science had been established and formulated, he would find the lectures largely incomprehensible. But as a classical scholar and grammarian who was also a physician (his task of ministering to the physical needs of, among others, Henry VIII and Cardinal Wolsey can have been no sinecure), Linacre doubtless had a good understanding of the interdependent relationships between the various areas of human knowledge, even within the relatively narrow frontiers which that knowledge had reached in his lifetime. Since Linacre's death nearly five centuries ago, those frontiers have expanded exponentially and their advance is still accelerating: but the interdependence, which Linacre would have recognized and which these lectures demonstrate so well, remains.

Linacre College's purpose in establishing these lectures has been to use the platform which a leading university provides to describe, for a much wider audience, the realities which should inform the environmental debate. Oxford is a particularly appropriate platform, since Oxford University has demonstrated its own commitment to environmental studies by establishing the new Environmental Change Unit, with the help of initial funding from IBM.

Bad news sells newspapers and swells television audiences, good news does not. It was therefore inevitable that the media, over the past decade, should have given greater emphasis to the dramatic aspects of environmental change—global warming, ozone holes, rising oceans, a nuclear winter—than to solid and sometimes difficult scientific evidence which puts these and other environmental problems into a proper perspective. For the same reason, the possibilities of positive corrective action have commanded less space and attention than doomsday scenarios which both feed on and fuel public concern. The purpose of these lectures, therefore, is to establish a sound and more balanced perspective on environmental change, based on the best available expert evidence. The lectures happened to be delivered during a dramatically eventful period, between October 1990 and March 1991. Britain acquired a new Prime Minister and one of our lecturers, Michael Heseltine, became Secretary of State for the Environment as a result. The crisis in the Gulf, following Iraq's occupation of Kuwait, exploded into full-scale war which, especially after Iraq's vengeful firing of Kuwaiti oil-wells and deliberate discharge of oil into Gulf waters, created new and formidable environmental hazards for the region—on which another of our lecturers, John Woods, was asked to advise.

The theme of the interdependence of factors bearing on the environment, exemplified by the broad scope of Richard Southwood's magisterial inaugural lecture, recurs frequently in the lectures which follow. John Mason refers to the sensitivity of climatic model results to a wide range of interacting processes, including cloud characteristics, fluxes of heat and moisture between the oceans and the atmosphere, feedback between changes in ice cover and ocean or atmospheric temperature, and changes between melting snow cover and soil moisture. Given this daunting range of variables, even the most advanced available computers can produce climatic predictions only within rather wide parameters of probability. James Lovelock elevates interdependence into a self-sufficient scientific view of our planet, the Gaia theory, which sees the earth as a self-regulating ecosystem. Interdependence emerges in a different context in Ghillean Prance's account of the dependence of Brazil-nut trees for pollination on particular species of large bee whose males are, in turn, dependent on certain varieties of orchid for the scent with which to attract females for mating. Such chains of dependence appear repeatedly in the lectures.

Another recurring theme in the lectures is that of the key role which the applied sciences can play in unravelling and, to an extent, solving the riddles of the environment. The imperfections of the climate models which John Mason describes will be significantly reduced, even if they can never be completely eliminated, by the development of more powerful computers. Given the rate at which computer science is advancing, it cannot be many years before even the CRAY YMP supercomputer becomes obsolete. More sharply focused predictions of greenhouse warming, and consequently of rainfall patterns and ocean levels will then be possible. Perhaps the most dramatic demonstration of the environmental importance of applied science comes from John Woods, who describes in detail the role of satellites, acoustic remote sensors and autosubs in making possible the dramatic breakthrough in oceanography which is now taking place. And James Lovelock points to the ways in which industry, with its technological resources, can be the white knight of the environment rather than the destructive demon of crude 'green' propaganda.

The capacities of applied science and technology for environmental analysis and protection will nevertheless be insignificant unless there is sufficient human and political will to use them and to learn from the results. Probably the most striking message of this first series of Linacre Lectures concerns the extent to which this will already exists and the ways in which it can be strengthened further. Richard Southwood illustrates the momentum which has built up over the past twenty years behind both national and international environmental legislation and stresses the role of individual everyday decisions: 'the prospects for the environment are in our individual hands'. (I should note in passing that this lecture inspired the students of Linacre College to form an Environmental Working Party to monitor college policy on purchasing, energy use, and other environment-related matters; this inevitably gives rise to some administrative headaches but is the kind of initiative which deserves to be imitated in all human communities, however small.) Michael Heseltine, in the wider context of government policy, presents a highly encouraging prospectus with a rightly international emphasis: 'The winds and the water pay no heed to national frontiers and in the context of environmental pollution, neither should we'. Crispin Tickell, speaking from professional experience as well as personal conviction, takes a similarly positive, activist approach; and John Phillipson makes telling points about the role of the individual in promoting nature conservation.

Although the message from these eight lectures is largely positive, not one of the lecturers could be accused of complacency. All of them—including James Lovelock, despite his spirited denial of the alleged 'fragility' of the Earth—agree that problems exist, both physical and behavioural, and that they are massively formidable. But to an increasing extent the problems can be identified and quantified; moreover, tools of ever greater sophistication are being put into our hands with which to tackle them. As John Mason puts it: 'It would appear that we have a breathing space of some 50 years but this may prove too optimistic; in any case, it is none too long'. These lectures may, we hope, make a small contribution to ensuring that the breathing space is sensibly and vigorously used.

1
The environment: problems and prospects
Richard Southwood

Sir Richard Southwood, FRS, Vice-Chancellor of Oxford University and Linacre Professor of Zoology, is one of Britain's foremost environmentalists. After nine years as a Lecturer in Zoology at Imperial College London and three as Reader in Insect Ecology at the University of London, Sir Richard was in 1967 appointed University Professor of Zoology and Applied Entomology while also heading that Department at Imperial College and directing the Field Station of the College. In 1974 he became Chairman of the Division of Life Sciences at Imperial College and in the same year was appointed to membership of the Royal Commission on Environmental Pollution (RCEP); he was Chairman of the Commission from 1981 to 1986. Meanwhile, in 1979, Sir Richard had moved to Oxford on his appointment to the Linacre Professorship of Zoology and his election to a Fellowship at Merton College. He has been a member of the National Radiological Protection Board since 1980 and its Chairman since 1985. He was installed as Vice-Chancellor of the University in October, 1989. He is the author of numerous books and papers on insects and ecology. Sir Richard's acceptance of Linacre's invitation to deliver the inaugural Linacre Lecture gave the series the most auspicious possible start.

Our environment is broadly defined as our surroundings or, as Einstein put it: 'Everything that isn't me.' In this broadest of definitions, you are part of my environment and I am part of yours. However when the word is used, as it is so often today, in relation to popular concern, it is the natural world that is in the mind of the speaker—the loss of the rainforests, the effect of acid rain on rivers or lakes, or the all-pervading effect of global warming. These and many other problems are recognized as being due to human activities, or to be 'anthropogenic' (to give a simple idea classical dignity). So although we may not immediately think of ourselves and our fellow human beings as part of the environmental picture, we are essential to its consideration. In the more than 3000 million year history of life on earth, undoubtedly the most striking biotic change was that brought about by the early photobionts, the first photosynthesizing microbes that produced oxygen; they would have been viewed as polluters by other extant organisms—if they could have expressed an opinion. In due course these changes were to drive the anaerobic organisms that originally populated the earth to take refuge in unusual environments, such as sulphur streams or the guts of other animals, or in most cases in extinction, though these events occurred over aeons of time by our standards. All organisms alter their environment to a greater or lesser extent, but in terms of speed and scale man's impact is now without precedent. Monitoring these changes is the theme of these lectures.

MAN'S IMPACT ON THE ENVIRONMENT

Human impact on the environment (I) is basically related to three factors: population size or the number of people (P), the amount of energy used per capita (E), and the extent that the energy used is non-renewable or leads to non-reversible degradation of the environment (N). These relationships may be simply expressed as (Ehrlich and Holdren 1972; Southwood 1972):

$$I = (P \times E) + (P \times E \times N)$$

Population

Whilst a couple of decades ago there was much public discussion of the difficulties posed by the burgeoning world population, a growth in part at least released by the improved control of disease in the tropics and sub-tropics, this aspect of the 'problem' is now much less debated. There has not been any drop in the rate of population growth to justify this—rather, world population continues to grow at a rate that exceeds predictions (World Resources Institute, International Institute for Environment and Development, and United Nations Environment Programme 1988), including those calculated on the same basis as that used to calculate the maximum rate for any animal population (Fig.1.1). Undoubtedly, high rates are related to poverty; a significant rise in the standard of living often leads to a marked reduction in the birth-rate. This phenomenon has

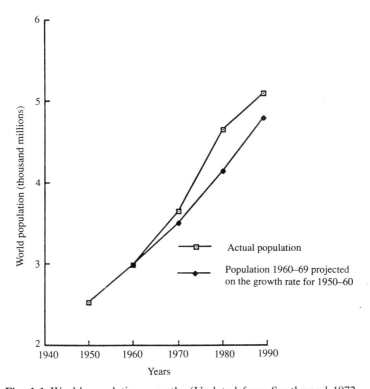

Fig. 1.1 World population growth. (Updated from Southwood 1972 and reproduced with permission from *Biologist—Journal of the Institute of Biology*.)

been called the demographic transition (Ehrlich and Ehrlich 1990). One reason why many have felt it difficult to press population restraint on the poorer countries of the world is the difference in per capita energy use. As is clear from the expression above, the impact is related to the product of population and energy use. It is quite clear that the more numerous, poorer countries use much less energy per capita (Table 1.1):

Table 1.1 Energy consumption per capita for countries grouped into three income categories, with examples from each category (kilograms of oil equivalent).

Country	1965	1988
High-income economies	*3707*	*5098*
United States of America	6535	7655
Australia	3287	5157
United Kingdom	3481	3756
Germany, Fed. Rep.	3197	4421
Middle-income economies	*585*	*1086*
Philippines	160	244
Mauritius	160	402
Brazil	286	813
Libya	222	2719
Low-income economies	*126*	*322*
India	100	211
Indonesia	91	229
Nigeria	34	150

(After *World Bank World Development Report 1990*)

one might argue simplistically that at present levels of energy use, every child born in the USA has 72 times the impact on the environment of a baby in India or 200 times that of the hunter/gatherer child.

Energy

During the last two decades there has been much discussion of 'energy conservation' and in some countries there have been modest policies directed towards this end. But despite this concern, and the effects of market forces when fuel prices were high during and shortly after 1973, energy consumption per capita in 1988 (the last available data) is substantially more than in 1965 (Table 1.1). Increases in consumption have occurred in virtually every country, proportionally more in most of the poorer countries, though even in the United States consumption grew (Fig. 1.2). There is a general relationship between energy consumption and measures of national prosperity, such as Gross National Product (Royal Commission on Environmental Pollution (RCEP) 1976*a*); the relatively modest increase in the UK might be thought to reflect the economic performance of this country, rather than any real success of energy conservation measures. Many other factors can be identified as contributing to the pattern of energy use, for example, climate (high

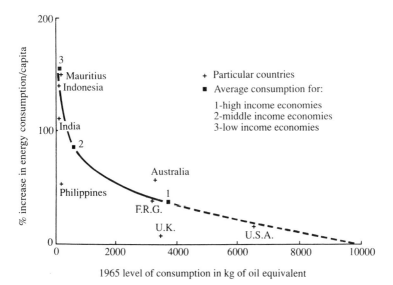

Fig. 1.2 Growth in energy consumption per capita against level in consumption in 1965 (the other countries in Table 1.1, i.e. Libya, Brazil, and Nigeria, are significantly above the line). (Based on data in Davis, *Scientific American*, 1990.)

in Canada with its cold winters) and availability (high in oil-rich states), but so far as environmental impact is concerned the conclusions from Fig. 1.2 are not encouraging. This shows that energy consumption per capita has continued to rise everywhere. If we are so bold as to project the graph and assume the continuation of present factors (leaving aside the argument as to what they are) stability will only be achieved at a staggering 9750 kg oil equivalent per capita. Recognizing the need for the poorer countries to increase their energy consumption, one might postulate that an ideal form of Fig. 1.2 would lack the inflection between the middle-income and richer countries so that the horizontal axis would be intersected at around 3000 kg oil equivalent. That is, we should see the demands of the richer countries falling; for at even this value the total world consumption would rise, because the vast majority of the world population are, at present, poor. The challenge would then be to cover this more modest total with renewable sources. This in itself would be formidable enough, for renewable resources still make only a very modest contribution to our requirements (Roberts *et al.* 1990).

The extreme modesty of progress in harnessing renewable resources must be a cause for concern; most economic analyses show that in the short term they are expensive and this has resulted in a paucity of investment for their development. Financial concerns, mostly associated with the costs of decommissioning, and anxieties over safety have reduced the nuclear power programmes of most countries, whilst fusion power is still a long way off. We are left with a heavy reliance on fossil fuels, undoubtedly the cheapest in the short term, but with long-term environmental effects discussed later in this lecture. There are also political costs of our dependence on resources where these are controlled by relatively few countries, as we saw during the Gulf war, but this aspect is beyond the scope of this lecture.

Looking to the future, many means of reducing energy consumption can be identified. The developments in electronics should lead to a great reduction in energy used for mechanical purposes, and their impact on communications should reduce the need for travel. If dreams of superconducting cables can be realized, then electricity can be moved around the world in phase with the daily peaks of demand; this would greatly improve overall efficiency, which is depressed by the need to maintain off-peak generation levels to cope with peak hour demands. However, progress is by no means consistent; improved building design can reduce heat loss but any gains in Britain in this respect seem likely to be lost by the growing tendency to install air-conditioning in modern offices. From the environmental viewpoint this must be regarded as a very retrograde development. Our architect colleagues should be able to gain more prestige from a design that utilizes natural air flows and takes account of solar aspect, thus saving energy and incidentally reducing the risks of Legionnaires' Disease and other infections. This presents us with a paradox. Cheap energy is an important contributor to a prosperous economy and in such an economy the environment is more likely to be a matter of concern than in a poor one (compare the former West and East Germany), yet as long as energy is cheap, and market forces reign alone, it will be used profligately and this will harm the global environment.

Food

With every organism but man, food represents its total energy demand, calculated from the energy content of the food itself and derived, directly or indirectly, from sunlight. For man not only is there the addition of the total energy used for heating, transport, and so on as discussed above, but his modern agricultural techniques often require an energy subsidy. Thus the amount of energy in a piece of pork is merely one-tenth of the energy utilized in its production. Artificial foods, fertilizers, and pesticides all require much energy for their production; further energy is used by farm machinery and in operations such as grain-drying. Methods of food production that have this energy subsidy also contribute to general global problems, but some of the more significant effects of agriculture on the environment are due to local pollution. Whereas an adequate daily food intake is thought to average rather less than 3000 kcals/day (there are differences due to age, stature, and sex), in many countries not only is production in excess of demand, but consumption is greater than the optimum for health (Table 1.2). The picture presented by Table 1.2 is, however, more encouraging than when I reviewed the data from the same countries in 1972 (Southwood 1972) in that all these particular underdeveloped countries now show modest increases in the amount of food eaten, whilst there is no marked increase in the consumption levels of the richer countries. At present, following the 'Green Revolution', there is no absolute shortage of food in the world, though many cannot afford to purchase the food grown with extra energy inputs, because of the addition to its price of the costs of this energy.

There are two environmental costs to high levels of food production. In the first place there are the pollution effects of the extra inputs. Secondly, these high levels may lead to degradation of the land, by soil erosion or overgrazing, though these problems are by no means confined to modern agriculture. However we must remember that one of the unique characteristics of man is the extent to which, by our endeavours, we have increased the carrying capacity of our environment. Carrying

Table 1.2 Estimated calorie content of national average food supply (in kilocalories/capita/day).

Country	1966–68	1969–71	1975–77	1978–80
Australia	3218	3300	3300	3190
Portugal	2915	3087	3135	3210
United Kingdom	3326	3352	3250	3326
United States	3380	3458	3564	3643
Brazil	2486	2486	2486	2510
India	1856	1989	1900	1989
Philippines	1890	1966	2124	2305
Libya	2313	2384	3186	3422
Mauritius	2384	2429	2610	2701

(Derived from FAO 1981)

capacity is an ecological term defined as the number of individuals than can be sustained on a certain amount of habitat, in our case, an area of land. Through agriculture, human endeavour has gradually increased the carrying capacity of the land over the last 4000 years, but most dramatically in the last few decades (Table 1.3). There is no way the world could completely turn its back on modern agriculture; there are just too many of us to feed, even if we were willing to work as medieval farm labourers did. A 'back to nature' strategy will not solve our problems. What we need to do is to reduce the term N (non-reversible degradation) in the equation cited above. We may modify nature, as indeed Table 1.3 shows we need to; but we must not reduce the carrying capacity of our environment, the earth.

Table 1.3 Carrying capacity for man.

Mode of agriculture	Number per square mile
Hunter-gatherer, unmodified tropical rainforest	1
Shifting cultivation, S.E. Asia	15
Medieval agriculture (c. AD 1200), UK	45–55
Intense tropical peasant agriculture, Tsembega, New Guinea	125
Modern agriculture, UK	400

(After Southwood (1975))

Pollution

Carrying capacity may be reduced temporarily or permanently by pollution, which may be defined as: 'the introduction by man into the environment of substances or energy liable to cause hazards to human health, harm to living resources and ecological systems, damage to structures or amenity, or interference with legitimate uses of the environment' (Holdgate 1979). By including amenity, this definition goes wider than the effects that will lower the carrying capacity and includes noise, which for many people is a major local concern. One man's music is another man's noise. However, as so defined, pollution is restricted to human activities, though many organisms produce substances that change ecological systems; whether these effects are regarded as harmful depends on the viewpoint. One may view the leaves that drop into a pond as pollution, and if they are very numerous in relation to the volume of water, the micro-organisms that develop on them may take up all the oxygen in the water causing fish deaths comparable to those due to pollution by sewage. In practice, if leaves drop into a pond they will do this every year and so there will not be a population of fish to be affected. To generalize: natural systems are adapted to and often buffered against any input from natural sources. They are resilient to natural perturbations; even bad episodes such as an outbreak of poisonous dinoflagellates, causing a 'red tide', do not make any lasting impression.

The resilience of a system depends on certain features of the population dynamics of organisms, principally their density-dependent response; that is, when they are scarce their rate of increase is greater. It also depends more globally on biogeochemical cycles. These are the cycles through which materials are circulated between organisms and the environment. One of the most important is the carbon cycle, since carbon is a key component of organic materials. The carbon dioxide that we breathe out is not a pollution problem because it is taken in by plants and made into plant material during photosynthesis, or it may be dissolved in the sea. This system has enough resilience to absorb initial increases in the output of carbon dioxide, such as occurred in the first stages of the industrial revolution. In other words there is a threshold below which increased output of a substance does not have any deleterious effect. At this level, which may be defined as that which does not overload the biogeochemical cycle, the substance is not a pollutant.

From this it follows that when considering pollution one should first divide substances into those that have such a threshold and those that do not. The latter may be described as pollutants, but substances that have thresholds can only be so described when above the concentration/time level of the threshold. For this reason I believe the original European Community approach of listing substances as pollutants, without reference to their concentration/time profiles, is mistaken. The threshold level may be lowered by the presence of another substance which is then known as a synergist.

The practical significance of this is that for those substances without a threshold one does not have to reduce their input to the environment to zero. A good example is crude oil: salt marshes actually benefit from a very small amount—it is broken down by micro-organisms and so contributes to the nutrient flow in the ecosystem (Royal Commission on Environmental Pollution 1984). One may draw an analogy between the process of pollution by substances that are part of natural cycles and the flow of traffic: as long as the quantity is not too great it flows smoothly, but once it increases above a certain level then it accumulates at certain bottlenecks.

Biocidal pollutants

These are pollutants whose major effects in the ecosystem are on the living organisms directly, rather than by their impact through the biogeochemical cycles. Often there is no threshold, and they are hazardous to the health of humans and/or other organisms. They may be divided into two main categories: those that are natural and those that are man-made.

Natural biocides These are materials that occur naturally, but man's activities increase their concentration in natural ecosystems. Often they are present mainly in the core of the earth, and man brings them to the surface and spreads them around into systems from which they have hitherto been absent. Two examples that are at present regarded as problems are lead and radiation: the latter is of course not a substance, but various forms of energy. Much of this radiation is quite independent of man's activities, though through these he may increase his exposure, but radioactivity also occurs in certain elements (e.g. uranium) that are mined.

Lead A decade ago there was widespread public anxiety about the effect of lead on the environment. There was particular concern about its effect on the behaviour and mentation of children. Although in some regions lead occurs in moderate concentrations in the soil, most of the lead in our immediate environment has been mined. Analysis of ice-cores from Greenland show how man's activities have distributed lead in the environment, with a slow increase from the start of the industrial revolution becoming a rapid rise since 1950 (Murozumi *et al.*, 1969). Lead is used for a variety of purposes, often in its metallic form, for example on roofs, as cable sheathing, or in batteries. These uses normally contribute little to the environmental burden of lead. So far as man is concerned this burden arises from three main sources: anti-knock compounds added to petrol, lead water pipes, and lead compounds added to paint. There is clear evidence that high levels of lead cause poisoning; some of the earliest evidence comes from ancient Rome where various drinks were warmed and stored in lead vessels (Gilfillan 1965). The clinical symptons of lead poisoning include anaemia, renal damage, tremor, and lack of co-ordination; these are normally exhibited if the blood lead level rises above 150 ug/dl,* but may sometimes be exhibited at levels as low as 50 ug/dl. But is there a threshold? Do these varied effects just cease at lower concentrations or are they expressed as more subtle effects, particularly in children, on the brain and other systems? Because of many confounding factors, attempts to determine the answer from surveys, such as those in which the amount of lead in the 'milk' teeth shed by children were correlated with measures of their intelligence, have invariably been somewhat inconclusive (Royal Commission on Environmental Pollution 1983). But lead is undoubtedly a pollutant. It remains where it arrives in the environment for a very long time: that is, it has a long environmental residence time (Table 1.4) and it is toxic to animal cells in all its compounds. It also has effects on microorganisms, and from its physiological effects on cells it is unlikely to have a threshold. So there can be no doubt that we should

*ug/dl = micrograms per decilitre.

Table 1.4 Environmental residence times for various pollutants.

Pollutant	Situation	Time	Percentage remaining
Carbon monoxide	1 m column of air in contact with soil	30 min	50%
2,4,5-T (herbicide)	Soil	Several weeks	50%
MCPA (herbicide)	Soil	Several days	50%
DDT (pesticide)	Soil	4 months	74%
Parathion (pesticide)	Soil	20 days	50%
Oil	Sea-water	4–5 weeks	70%
Radioactive iodine (^{131}I)	—	8 days	50%
Radioactive strontium (^{90}Sr)	—	28 years	50%
Cadmium	Upper layers of moorland soil	6–20 years	90%
Lead	Upper layers of moorland soil	70–200 years	90%

(*Source:* RCEP 1983)

seek to minimize the amount of lead in the environment. However, any action has its costs: in this case those of the elimination of anti-knock lead compounds from petrol, of lead compounds from paint, and the replacement of lead water-pipes, especially in areas where the water is somewhat acidic. The problem faced by the Royal Commission on Environmental Pollution (RCEP) when it was reporting on lead in the environment was how to make a choice between the costs of reducing pollution and those associated with an unquantifiable risk to human health. Most public attention was focused on the lead in petrol. This was justified from the general environmental viewpoint because of the quantity involved and its wide dispersion (Fig. 1.3). The manufacturers of the additives, aided and abetted by many car makers, produced very high estimates of the costs of eliminating these anti-knock additives. The Commission had to form a view on the validity of these estimates; if they were right the costs were indeed horrendous. My doubts

18 The environment: problems and prospects

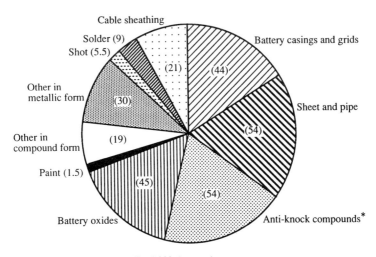

Figures in thousand tonnes. Total 283 thousand tonnes.
*This figure if for manufacture of anti-knock compounds, about 80% of which are exported.

Fig. 1.3 UK lead consumption in 1982. Figures given are for thousands of tonnes, out of a total of 283 thousand tonnes. (*Source*: RECP 1983.)

on these costs were aroused by the fact that the automobile industries of Japan, Sweden, and the USA (not noted for their uncompetitiveness) were able to manufacture cars that would run on lead-free petrol. A thorough analysis showed that there were no fundamental obstacles to lead-free petrol. Theoretically one would get slightly fewer miles per gallon, but many other aspects of car design were far more important in determining that factor. Furthermore, there would be many improvements in car efficiency in the next decade, so that any loss due to using lead-free petrol (that is a lower octane grade) would be a small future benefit foregone, rather than an extra cost. No one would find their fuel costs going up, they would merely not go down so quickly. Though this was important from the political, public-perception, viewpoint, it was still a real, though now much reduced, cost. The other major costs were in the manufacturing of fuel and cars. The oil companies readily confirmed that given time they could change over to the new fuel at virtually no cost: after all, they were already producing it for many countries. This provided the key, the time-scale of change. The cost of resolving an environmental problem often depends critically on the speed with which the change is brought about. Manufacturing machinery is frequently changed, and environmentally desirable modifications can be incorporated at these times at very little cost, but this will not happen unless it is known that there will be a mandatory requirement for these changes. I will return to this aspect later. Taking these considerations into account, the scientific and economic uncertainties of the problem were evaluated (Royal Commission on Environmental Pollution 1983; Southwood 1985*b*). The view might then have been taken—it often is in environmental matters—that we should not do anything until we are sure. But of course we are already doing something, in this case distributing lead in the environment. The question is then: 'what if this proves to be the wrong thing to do?' Therefore the next and final step is a managerial judgement of the balance of uncertainties. This can be quantified as follows:

$$x(a) \parallel 1 - x(b),$$

where x is the scientific guesstimate of the probability of a certain risk being real, a the costs of the risk, and b the costs of eliminating the risk. The two parallel lines $\parallel$ mean that the terms on the left are compared with those on the right—the decision whether or not to eliminate the risk rests on whichever side of the expression is the greater.

It is now history that the Government, under considerable pressure from the 'CLEAR' campaign, accepted the recommendations of the Royal Commission within hours of their publication (a record I believe for a Royal Commission). Lead additives were to be phased out of petrol and after 1990 all new cars were to be capable of running on lead-free petrol. Remembering the polemics at the time I have been fascinated to see how quietly this change has come about.

The two other major sources of lead for man were paint and, in some regions, drinking water. The paint manufacturers accepted a timetable for reducing the lead content of paint, which must now be indicated on the container. The problem of lead in drinking water remains; some progress has been made, but much remains to be achieved. It is difficult to get reliable comparative data on blood lead levels, but those for women in Wales do show a decline over recent years (Fig. 1.4).

The lead from petrol was widely distributed in the environment, but other human contamination was from more localized sources. The most serious lead-poisoning problems for other animals also arise from local sources; for example the lead poisoning of mute swans (*Cygnus olor*) on the Thames and other waterways caused by the swallowing of anglers' weights. This was thoroughly investigated by Dr C. M. Perrins and his colleagues in the Zoology Department at Oxford University (Birkhead 1982, 1983; Birkhead and Perrins 1985; Sears 1988). The effects on individual swans were distressing. In particular the inability to hold the neck straight led eventually to death,

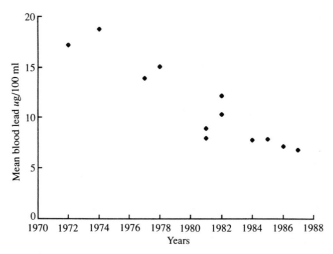

Fig. 1.4 Lead blood levels in women in Wales. (Data extracted from Elwood (1983) and Department of the Environment Blood Lead Monitoring Programme 1984–87 (1990).)

and the national population declined. Following the Royal Commission's recommendations on the basis of this research, regulations were made to phase out lead weights. This has now occurred and the swan population is recovering (Sears 1989).

There are other pollutants that fall into exactly the same category as lead: for example, mercury, titanium, chromium, and cobalt. We can draw general lessons from the history of lead pollution. Firstly, although there is no threshold, the problem only becomes significant when man's use becomes widespread (e.g. lead in petrol, anglers' weights). The solution to the problem lies in technological changes that prevent the material being scattered in the environment. When these are implemented, improvements occur.

Also in the same general category, that of natural biocides, are various compounds of carbon, ranging in complexity from carbon monoxide to complex molecules such as the dioxins. Unlike the heavy metals, these are mostly broken down in nature to harmless materials, often quite quickly. Hence the extent to which they are a problem depends on their concentration, and they are essentially local problems with technological solutions, for example the processing of waste through incinerators with the correct time–temperature relationships.

Radioactivity Radioactivity is both a natural and an ancient phenomenon. A significant amount arrives at the surface of the earth from cosmic sources, and until the early marine organisms had produced sufficient oxygen for the ozone shield to form, the doses of some forms of radiation at the surface of the earth would probably have been such as to prevent life emerging on to land. Life originated in the oceans about 3000 million years ago, and complex organisms were abundant over 600 million years ago, but it was not until the end of the Devonian (about 370 million years ago) that animals and plants first lived successfully on the land surface. The major part of natural ionizing-radiation now comes from the earth itself, from rocks and from the soil (Fig. 1.5). A high level of radiation is indeed a powerful biocide, but we must assume that existing organisms can not only survive existing levels of radiation, but do so without any loss of biological fitness. This does not mean that the effects of radiation may not shorten the life of those past reproduction; in this connection it is notable how many cancers are problems of later life. It is not always appreciated how much the dose from natural radiation varies depending on where one lives (Fig. 1.5), or how much one can alter this by one's lifestyle (Clarke and Southwood 1989). Nevertheless, there is no threshold and thus doses should be kept 'As Low As Reasonably Achievable' (the ALARA principle). Problems (in the sense of this lecture) with exposure to ionizing radiation occur in three ways. Firstly there is radon; for most people the inhalation of this gas and its short-lived decay products is the principal mechanism of exposure to radiation (O'Riordan 1990). Radon emerges from the ground, being produced in the natural decay of uranium. The quantity is very dependent on the nature of the underlying soil and so concentration in houses, where it may accumulate, varies greatly from one part of a country to another. In the UK, it is greatest in the south-west. Relatively simple measures will considerably reduce the concentration of radon in houses, whilst, as with so many health problems, smoking increases the risks of lung cancer from this cause (O'Riordan 1990; Clarke and O'Riordan 1990; Committee on Biological Effects of Ionising Radiation (BEIR IV) 1988). The greatest exposure to artificial (that is, man-produced) ionizing radiation generally occurs during medical treatments; clearly there are major benefits to set against the risks, but here it is just as important to avoid unnecessary exposure. Much is being done and more should be done to reduce patient exposure in diagnostic radiology; there is considerable potential to do so using modern equipment (Royal College of Radiologists and National Radiological Protection Board (RCR and NRPB) 1990). Last and least, but causing most concern, is the radiation arising directly or in-

Fig. 1.5 Variations of annual radiation dose in the UK compared with the average. Pie areas are proportional to dose, and the percentage segments are proportional to sources. (*Source*: Clarke and Southwood 1989. Reproduced by permission from *Nature*, **338**, 197–198. Copyright © 1989 Macmillan Magazines Ltd.)

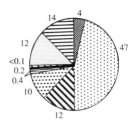

(a) Average person in UK, 2.5 mSv overall

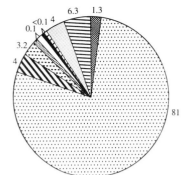

(b) Average person in Cornwall, 7.8 mSv overall

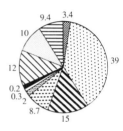

(c) Heavy consumer of Cumbrian seafood in 1987, 2.9 mSv overall

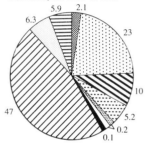

(d) Heavy consumer of Cumbrian seafood in 1983, 4.8 mSv overall

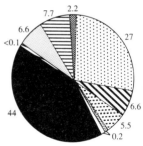

(e) Average nuclear industry worker in 1987, 4.5 mSv overall

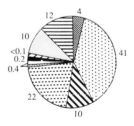

(f) Frequent air traveller, 2.9 mSv overall

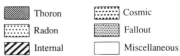
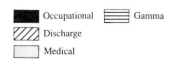

directly from the nuclear power programme and the processing of nuclear materials. The amount of radiation exposure that results is generally small, particularly in all recently constructed plants, yet concern remains high. The association of clusters of childhood leukaemia with certain nuclear plants has heightened anxiety (Jones and Southwood 1987; Gardner 1990a,1990b), though the general levels of radiation are so low that to cause such effects there must be some as yet unknown physiological mechanism leading to deposits of radionucleides and high spot doses to fetal haemopoietic tissues. Research continues, but no mechanism has yet been found (Morgan *et al.* 1990). Leukaemia cases are often clustered (Linet 1985), and the cause could well be other than radiation.

I do not believe that the normal operation of the recently constructed nuclear power plants poses any significant environmental problem, but two major problems remain. One is the safe disposal of the waste, a problem that concerned the Royal Commission in 1976 (Royal Commission on Environmental Pollution 1976b), and one that has remained unresolved, largely because of public perception of risk, rather than technical difficulties (Lee 1989; Southwood 1990b). The second problem is the risk of a nuclear accident. On the basis of calculations it has been predicted that accidents at nuclear plants should be extremely rare, but in fact we have had three. The last, at Chernobyl, was much the worst and its impact was more widespread than was initially realized. For example it was not predicted, when the cloud crossed the British coast, that radioactivity levels in hill-sheep in Wales would render the meat unacceptable for human consumption for a number of years (Wynne 1991). We are still learning about Chernobyl. We need to be confident that an accident of that magnitude is never again possible. If the public could be reassured, then nuclear power would have an important role in reducing the risk of global warming. A new concern is arising in relation to non-ionizing radiation from power lines and short-wave communication towers. At present there is considerable scientific uncertainty about its effects.

Man-made biocides The great majority of man-made biocides are used in agriculture as pesticides. They have been essential in the increase in the carrying capacity of the environment for man to which I have already referred. The environmental problems that arise do so because of their profligate use (Southwood 1980), a tendency encouraged by the drive for a higher level of productivity per acre (intensification) and by the consumer demand for perfect produce. The latter leads to what may be termed the 'cosmetic' use of pesticides, when they are applied not to increase the quantity of the crop, but to enhance its appearance and hence its value. The great price differentials between grades are such as to permit, on economic grounds, the application of a very large number of cosmetic applications of pesticide as an insurance against blemishes and a consequent drop in grade (RCEP 1979). Excessive use of pesticides damages the ecosystem, often destroying many non-pest species; it can lead to the development of resistance to the pesticide in the pest (demanding even higher levels of use or the loss of the pesticide as a method of control). The intensification of agriculture also causes a number of other environmental problems, and there are those who argue that it is irrational when it is causing over-production and food surpluses in Europe and North America. Alternative policies are being advocated: some propose organic farming but the labour costs are too high for the mass supply of food; and there is now much discussion of what is termed 'integrated' agriculture (the combination of natural methods with the precise use of pesticides and fertilizers) (Sansavini 1989). In tropical and developing countries the objective should be a sustainable system. As with European and American agriculture, the difficulty is not a lack of technical knowlege, but the pressures produced by the economic framework (Conway and Barbier 1990).

Pollutants that disturb the physico-chemical systems
Under this heading fall those major pollution problems that are described as 'acid rain', 'the ozone hole', and 'global warming'

(see Mason, Chapter 3). They all involve the modification of the complex chemical cycles that maintain our environment in its present conditions. The changes that result have caused (in the case of acid rain) or are likely to cause profound changes in natural ecosystems. They are large-scale international problems. There are other pollution problems in this category that, since they do not involve the atmospheric system, are not so widespread, for example ground-water pollution, and sewage pollution at sea.

'Acid rain' More correctly this should be described as acid deposition, for it refers to the deposition of acidic materials in rain, as fog, and directly from the air on to foliage. The 'acids' come from combustion of fossil fuels and are principally sulphuric and nitric. In order to implement the Clean Air Acts (1956, 1968) power stations in the UK were fitted with tall chimneys on the 'dilute and disperse principle' (Ashby and Anderson 1981); measurements of pollutants were made at ground level in the vicinity of the stations and pollution was shown to have fallen. However early in the 1970s it was noted that many streams and lakes in southern Scandinavia were losing their fish (principally trout and salmon) and this was associated with an increased acidity of the water, which was in turn related to increased acidity of the rainfall, due to the carriage of power station emissions across the North Sea on the prevailing south-westerly winds. Later some similar observations were made in Canada and in some parts of the USA. There, as in western Germany, and in Switzerland, coniferous trees were dying or showing signs of damage that were also attributed to the effects of 'acid rain'. Restricting this brief review simply to Britain and Scandinavia, the British authorities were unwilling to accept the Scandinavian account, which contained many assumptions. Alternative explanations were advanced: there had been extensive planting of conifers and it was known that when such plantations aged the soil became more acid. The methodology for determining the acidity of the rainfall could be questioned, particularly some of the early

records which were essential in establishing the trend of falling pH values. Others wondered why, if such effects occurred, they had not been detected in Britain, especially in Scotland where the soils are thin and acidic. It became a matter of controversy between governments. In 1983 the then Central Electricity Generating Board (CEGB) and the National Coal Board agreed to fund a research programme to be carried out under the auspices of the Royal Society, the Royal Swedish Academy, and the Norwegian Academy of Science and Letters (Mason 1990). It was known as the Surface Water Acidification Programme (SWAP). I was appointed Chairman of the Management Committee and Sir John Mason the Programme Director. It should be recorded that initially there was, in Scandinavia, considerable distrust of the motives of the programme. It was seen as a delaying tactic by Britain. However, multinational teams were formed and excellent collaborative work was undertaken in many areas, especially in palaeolimnology (Battarbee and Renberg 1990). After the interim conference in 1987 it was possible to report to the CEGB, and informally to the British Government, that there was no doubt that emissions from the UK were having effects in southern Scandinavia and thus some control measures were needed. When SWAP was established, there was the clear undertaking by the then chairman of the CEGB that its findings would be accepted and a programme of retrofitting flue-gas desulphurization equipment to the large Drax station was announced. The final report, based on a conference held in London in March 1990, confirmed that waters with impoverished fish populations receive high levels of acid deposition and there was agreement between the scientists on all the other major issues (Mason 1990; Southwood 1990*b*). The prime ministers of the three countries spoke at the final dinner and it was gratifying to observe how national differences could be resolved by such a scientific partnership. Is this a 'blueprint' for resolving such disputes? The organizational arrangements for this work were unique, but a key feature was the undertaking by Lord Marshall, on behalf of the CEGB, to accept the conclusions and act upon them. It is surprising that the British Treasury agreed to such a commitment, but there were certain Whitehall departments that were pressing for some accommodation of the Scandinavian complaints. Cynically, one can wonder if the Treasury was lulled into thinking that the scientific conclusion would go the other way! Today there are those who wonder if the previous policy will be accepted by, or imposed on, the electricity companies now they are being privatized and divided. Reduction in sulphur emission can of course be obtained by burning fuels that contain less sulphur, as well as by flue-gas desulphurization (Roberts *et al.* 1990). For such a policy to be really effective, the supply of the fuels must be assured. Both in Europe and North America acid rain is a pollution problem where, so far as freshwaters are concerned, there are now no real scientific uncertainties; but the costs and benefits are asymmetrically distributed between the countries involved. The 'polluter pays' principle is not easily applied in practice in an international context.

Ozone layer depletion The ozone in the stratosphere forms a shield against damaging incoming radiation; it is vital for the maintenance of life in terrestrial environments. Chlorofluorocarbons (CFCs) are synthetic chemical molecules that have been extensively used for about 40 years in refrigerators and as propellants in aerosols; they are stable and light. There is no natural sink for these molecules and in the mid-1970s atmospheric chemists expressed anxiety that they could rise to the stratosphere and there react with the ozone layer, thereby depleting it (Albritton 1989). However, there was much scientific uncertainty; as with so many of these problems considerable yearly variations make it difficult to be confident of the trend. This uncertainty is emphasized by those industries and countries engaged in the manufacture of CFCs. In the last decade, work in the Antarctic suggested that the ozone layer there might be thin; in the last three years this hole has been confirmed and it is now widely agreed that 'there is a real problem . . . given the will it can be solved' (Porter 1989). The Montreal Protocol attempts to provide that 'will' (Tolba 1989).

Global warming Certain gases in the atmosphere reflect back radiation from the earth; if the concentration of these rises, one might expect the temperature of the earth to increase (Table 1.5). Initially emphasis has been placed on carbon dioxide (Leggett 1990), there having been a great increase in the amount released by the burning of fossil fuels. Records of the amount in the air kept at the Mauna Loa observatory in Hawaii since 1958 have also shown a steady increase, though the increase is less than that expected from the quantity released from burning fossil fuels. It is thought that the difference is due to absorption in the oceans. Another factor that will contribute to the accumulation of carbon dioxide is the destruction of the rainforests (Myers 1990); a fully mature forest is in a steady state in its carbon budget, but when it is felled and burned, carbon dioxide is released to the air. Reforestation would provide a sink for, that is take up, carbon dioxide until the first generation of trees had matured. However it is now recognized that many other gases that reflect back radiation are also increasing, especially methane and the chlorofluorocarbons, a molecule of the latter having an effect many thousand times that of a molecule of carbon dioxide (Table 1.5); these gases are expected to contribute half as much again as the doubling of carbon dioxide (Mason 1989). Complex computer

Table 1.5 Global warming potentials for various greenhouse gases on molar and weight bases relative to carbon dioxide.

Gas	Residence time (*Years*)	Global warming potential	
		(*Molar basis*)	(*Weight basis*)
CO_2	230.0	1.0	1.0
CO	2.1	1.4	2.2
CH_4	14.4	3.7	10.0
N_2O	160.0	180.0	180.0
HCFC-22	15.0	810.0	410.0
CFC-11	60.0	4000.0	1300.0
CFC-12	120.0	10000.0	3700.0

Adapted from Lashof and Ahuja (1990), by permission from *Nature*, **344**, 529–31.

models have been developed to simulate the changes, but there are still many uncertainties and it seems doubtful if there is the basic information to improve the models greatly. All the models predict a temperature increase, its magnitude varying between 1°C and 5°C, with the doubling of the carbon dioxide level that will occur in the next 50–130 years. The changes in global rainfall and storm patterns are likely to be more significant for agriculture and ecosystems than the temperature change, though this is occurring at a greater rate than at any time in the last 12 000, perhaps 100 000, years (Southwood 1985a). A rise in sea level is also likely, though its magnitude is uncertain (Mason 1989). The mitigation of these effects clearly requires international collaboration on a truly global scale.

PROSPECTS FOR CONTROL

Attitudes

Over the last two decades there has been growing public appreciation of environmental problems. This has been particularly marked in the most prosperous countries. Concerns about the environment appear to take third place in a hierarchy of anxiety, behind wars and economic problems. It is therefore not surprising that some of the most vigorous 'environmental' movements have developed in California and Germany. Political parties, the 'Greens', espousing environmental causes have received significant support in many European countries, doing particularly well in Britain in the last elections to the European Parliament. Although many committed environmentalists feel the Greens oversimplify the solutions to problems, their presence has spurred all major political parties in Britain to give some attention to environmental issues in the programmes they have presented at recent elections.

Popular concern may manifest itself in the NIMBY (Not In My Back Yard) syndrome (Royal Commission on Environmental Pollution 1984), resulting in objections by members of the public to various potentially polluting activities being sited in the vicinity of their homes; this is a particularly difficult problem in the development of a rational waste-disposal policy (Royal Commission on Environmental Pollution 1985). A more ethical approach is NIABY (Not In Anyone's Back Yard), certain technologies being regarded as unacceptable anywhere. Many people hold such a view in respect of nuclear energy, more particularly after the Chernobyl accident, though nuclear power would provide an important component in any strategy to eliminate the risks from global warming.

Industry and commerce in the developed countries have responded to this growth in public concern, some developing particular 'environmentalist business management' schemes (Winter 1988). Public pressure in developed countries has influenced the activities of multinationals in the Third World. The World Bank now includes environmental aspects in its assessments.

Although on certain issues some countries and many companies take refuge in scientific uncertainty to avoid action, there is a growing willingness to recognize environmental problems. However, their solution can be both difficult and expensive; they place a constraint on the free rein of market forces (so popular with the world's major politicians) and the country or company that follows a strict, environmentally sound policy may find it a competitive disadvantage. Once an environmental problem has occurred, retrospective solution is usually expensive and often difficult to achieve. Industries that led the way in environmental matters, as the Japanese car industry has done with lead-free petrol, now have the advantage. Over the last two decades, largely since the United Nations Conference on the Human Environment in Stockholm in 1972, a number of concepts and mechanisms for the formulation and the implementation of policy have been developed.

Concepts

These may be placed in three groups:

(1) those that relate to the environment;

(2) those that relate to man;

(3) those that relate to economic factors.

In practice the three components must be brought together in any analysis; this occurs in the German principle of *Vorsorgeprinzip* and the British one of BPEO (Best Practicable Environmental Option). As the Royal Commission on Environmental Pollution (RCEP 1988) points out, BPEO is best considered in the reverse order, that is, option, environmental, practicable, best. The basis of the concept, first put forward by the Royal Commission in its fifth report (RCEP 1976*a*), and subsequently elaborated in its tenth, eleventh and twelfth (the latter being devoted to it) is that there may often be an option as to the medium (air, land, or water) into which a pollutant is discharged. One must then assess the environmental damage likely to result from each option, and then the costs of each. The level of costs provides a measure of the practicability of each option. Finally an assessment is made of the balance between the environmental benefits of each option and the costs of achieving it.

Rather more restricted are the concepts of ALARA (As Low As Reasonably Achievable) and ALATA (As Low As Technically Achievable). These refer to the quantity of a pollutant released into the environment. In ALARA the costs of reductions in amount of pollutant released are balanced against the benefits, generally measured in terms of reduction in risk. ALATA on the other hand disregards considerations of cost, and the levels of reduction are limited only by technical feasibility. As this is rather impractical it has been replaced by BATNEEC (Best Available Technology Not Entailing Excessive Costs), though one still has to make an inevitably subjective assessment of 'excessive' in relation to the acceptability of the risk. The level of risk to human life that societies are willing to 'accept' or 'tolerate' is itself a complex issue; purely mathematical measures of probability are seldom adequate. The extent to which a risk is known and the degree of dread are the main modifiers (Royal Society 1983; Southwood 1990*a*). Two other concepts, focused on economic considerations, are or have been applied in Britain: they are the 'polluter pays principle' and the 'Best Environmental Timetable' (BET). The former attaches the costs of any pollution to the process or product. The second, advanced by the Royal Commission in its ninth and tenth reports (Royal Commission on Environmental Pollution 1983, 1984), extends the timetable for pollution reduction so that its costs become sufficiently low to justify the measure. As already mentioned, the Commission argued that the costs of eliminating lead from petrol would be much reduced if the automobile industry was given seven years in which to adapt. Likewise it recommended that a policy to ban straw-burning at a certain date in the future should be established, thereby providing the incentive for the investment in machinery for alternative uses.

The German concept of *Vorsorgeprinzip* is not easily summarized (Royal Commission on Environmental Pollution 1988): it is concerned with the prevention and step-by-step reduction in environmental pollution. It requires the minimization of risk to environmental quality and the implementation of ecological principles.

MECHANISMS FOR CONTROL

International and national initiatives

The formulation of policy may be stimulated by conferences, either major international ones, such as Stockholm in June 1972, or more limited scientific meetings, for example the Bergen meeting of SWAP (Surface Water Acidification Programme) in June 1987. In practice, as major decisions usually depend on politicians, the most effective stimulus is a media-led public outcry. Those expert in a field will often feel that the media coverage of a topic is not balanced, that a relatively small risk is given a high priority; an example is provided by the attention paid to the Sellafield reprocessing plant compared with other risks from radiation (Clarke and Southwood 1989; Southwood 1990*a*). Nevertheless, without the public pressure generated in this way, environmental concerns would be unlikely to prevail. One of the most terrifying proposals for environmental modification was that developed in the 1970s to reverse the flow of rivers in north-central USSR. Arguments against were put forward by many scientific bodies, but it was not until the advent to power of Mr Gorbachev in 1985 that the press began to publish articles against the implementation of the project. After more than a decade of 'behind the scenes' discussion, but little more than a year after the first newspaper article, the Council of Ministers adopted a resolution cancelling the project (Tarnavskii 1990).

Precise recommendations arise from expert committees or commissions (Everest 1989) and they may be embodied in national law and in international agreements, the latter often sponsored by the United Nations or other international agencies. In the last decade in western Europe, directives from the EC Commission have proved a powerful driving force for environmental legislation. Because of the costs of most such measures, it is often most convenient for national governments to postpone action; this is exemplified in Britain by the continued delays in the full implementation of the provisions of the Control of Pollution Act (1974) (Royal Commission on Environmental Pollution 1984). International agreements may be rendered ineffectual by the refusal of powerful countries to adopt them, and bring in the necessary national legislation. International law being consensual, it cannot be created without the support of the powerful nations (Guruswamy 1990). The rejection of the United Nations Convention of the Law of the Sea (UNCLOS) by the USA and most western European nations has robbed that long-negotiated agreement of much of its effectiveness.

There are certain interesting differences between environmental legislation in the USA and that in western Europe (King 1989). These were especially well marked about a decade ago when US regulators were more aggressive, and decisions on risk were characterized as being confrontational, formal, and open to public participation (Jasanoff 1986, 1990). The resulting procedures were dominated by lawyers and hence extremely costly; they also led to courts making judgements on complex matters of scientific uncertainty. There was no necessary consistency between different decisions. The approach in western Europe, although revealing some national differences, was much more consensual, cost-conscious, and largely closed to the public. Both approaches have their disadvantages and are now less distinctive but, as Jasanoff (1990) has pointed out, because of the variety of scientific advice publicly expounded in the USA ('scientific pluralism'), public and political judgement will always play an important part in risk decisions in the USA.

Turning to the major environmental problem facing the world, global warming, on the positive side one can point to the recognition of the problem by leading politicians, even those with a strong market-forces creed. The United States was a leading protagonist of the Montreal Protocol, an international agreement to regulate the production of CFCs. Recognition of the problem has occurred, but will this be carried through to effective international action? The auguries are not good; the example of the failure of the major industrial nations to implement the Law of the Sea (UNCLOS), the attitudes currently being taken by the President of the USA on global warming, and the fate in Britain of the proposal for a 'carbon tax' all signal a current reluctance to enact effective policies, which will inevitably have costs (Guruswamy 1990). It may be argued that at the present time* the world economy is being stressed by the change to a market economy in eastern Europe and the USSR and by the 'Gulf crisis' and that therefore politicians cannot face a further challenge which could lead to a fall in living standards and in their electoral popularity! Ironically, to the extent that rises in oil prices encourage energy conservation measures, they are a plus for environmental protection.

*The lecture was delivered on 11 October 1990.

The role of individuals

Undoubtedly the most important force for the resolution of environmental problems in the future is public concern. This can influence the outcome through three mechanisms:

(1) Political pressure;
(2) Legal steps;
(3) Individual actions.

Throughout western Europe the major practical effect of the Green parties has been to ensure that all parties paid attention to environmental issues, often leading to positive action by the party in government. It is only realistic to recognize that in the absence of a disaster, environmental protection measures are viewed by most people as analogous to an insurance policy that can be justified only when the economic conditions are so buoyant as to permit the resultant expenditure to be painless. It is therefore not a matter of blaming the politicians, but of convincing the general public that the 'insurance' is worthwhile, even if the economic effects are not quite painless.

Environmental law is a new, but rapidly developing area (Kramer 1989). It can provide a useful means whereby the concerned citizen can exercise influence; this is particularly true in the legal system of the USA. In Britain legislation has often been drawn up specifically to exclude private prosecutions of public bodies on environmental matters, but as society becomes more concerned with these issues such conventions are likely to change.

The third and perhaps the most effective route for individuals is through their own decisions. At the beginning of the lecture I stressed that the impact of human activities on the environment was related to population size, energy use per capita, and the extent of non-sustainable activities. Thus it follows that the prospects for the solution of environmental problems rest on the myriad of decisions that we all individually make; the speed at which we drive, the type of transport we use, the products we buy, our demand for blemish-free fruit, the temperature we maintain in our homes and offices, and even such simple actions as turning out lights when not needed. Through all these, however small, we influence the course of global events.

The extent to which consumer power has caused manufacturers to develop 'environmentally friendly' products is most encouraging, though some advertised claims are spurious and should attract the attention of the appropriate regulatory authorities. The Consumers Association has produced literature guiding the individual to 'environmentally-sound' actions, and a variety of other groups (Friends of the Earth, National Society for Clean Air, etc.) provide abundant information and publicity on particular issues. On the other hand, it is profoundly discouraging that air-conditioning seems to have become regarded as a standard for new offices in the UK, without any consideration, so far as I can see, of the environmental disbenefits. If we continue to follow America in these energy-profligate fashions we will continue to increase our per capita energy demands with dire consequences.

The key feature of our strategy for environmental protection in the future must be an integrated view of all our activities and their environmental consequences (Guruswamy 1989). The proposal in the recent White Paper to make all relevant departments consider the environmental implications of their actions is an important signal in this respect. Science can identify the problems and generally discover the ways for their solution, but the prospects for the environment are in our individual hands.

REFERENCES

Albritton, D. L. (1989). Stratospheric ozone depletion: global processes. In *Ozone depletion, greenhouse gases, and climate change*, pp.10–18. National Academy Press, Washington, DC.

Ashby, E. and Anderson, M. (1981). *The politics of clean air*. Clarendon Press, Oxford.

Battarbee, R. W. and Renberg, I. (1990). The Surface Water Acidification Project (SWAP) Palaeolimnology Programme. *Philosophical Transactions of the Royal Society of London*, B, **327**, 227–32.

Birkhead, M. (1982). Causes of mortality in the Mute swan *Cygnus olor* on the River Thames. *Journal of the Zoological Society of London*, **198**, 15–25.

Birkhead, M. (1983). Lead levels in the blood of Mute swans *Cygnus olor* on the River Thames. *Journal of the Zoological Society of London*, **199**, 59–73.

Birkhead, M. and Perrins, C. M. (1985). Breeding biology of the Mute swan *Cygnus olor*. *Biological Conservation*, **32**, 1–11.

Clarke, R. and O'Riordan, M. (1990). Rumours of radon. *Science and Public Affairs*, **5**, 25–36.

Clarke, R. H. and Southwood, T. R. E. (1989). Risks from ionising radiation. *Nature*, **338**, 197–8.

Committee on Biological Effects of Ionising Radiation (BEIR IV) (1988). *Health risks of radon and other internally deposited alpha-emitters*. National Academy Press, Washington, DC.

Conway, G. R. and Barbier, E. B. (1990). *After the green revolution: sustainable agriculture for development*. Earthscan Publications, London.

Davis, G. R. (1990). Energy for planet earth. *Scientific American*, **263**(3), 21–7.

Department of the Environment UK Blood Level Monitoring Programme 1984–87. (1990). Results from 1987. *Pollution report no. 28*. HMSO, London.

Ehrlich, P. R. and Ehrlich, A. H. (1990). *The population explosion*. Simon and Schuster, New York.

Ehrlich, P. R. and Holdren, J. P. (1972). One dimensional ecology. *Science and Public Affairs: the Bulletin of the Atomic Scientists* (May 1972).

Elwood, P. C. (1983). Changes in blood lead concentrations in women in Wales, 1972–82. *British Medical Journal*, **286**, 1553–5.

Everest, D. A. (1989). *The provision of expert advice to government on environmental matters: the role of advisory committees*. (Research report no. 6). Environmental Risk Assessment Unit, UEA, Norwich.

Food and Agriculture Organisation (FAO) (1981). *The state of food and agriculture*, (FAO Agriculture series; no.14). FAO, Rome.

Gardner, M. J. (1990*a*). Results of a case control study of leukaemias and lymphomas among young people near Sellafield Nuclear Power Plant in West Cumbria. *British Medical Journal*, **300**, 423–9.

Gardner, M. J. (1990b). Methods and basic data of a case control study of leukaemias and lymphomas among young people near Sellafield Nuclear Power Plant in West Cumbria. *British Medical Journal*, **300**, 429–34.

Gilfillan, S. C. (1965). Lead poisoning and the fall of Rome. *Journal of Occupational Medicine*, **7**, 53–60.

Guruswamy, L. D. (1989). Integrating thoughtways: re-opening of the environmental mind? *Wisconsin Law Review*, **3**, 463–537.

Guruswamy, L. D. (1990). Global warming: integrating United States and international Law. *Arizona Law Review*, **32**, 221–278.

Holdgate, M. W. (1979). *A perspective of environmental pollution*. Cambridge University Press.

Jasanoff, S. (1986). *Risk management and political culture*. Russell Sage Foundation, New York.

Jasanoff, S. (1990). American exceptionalism and the political acknowledgement of risk. *Daedalus, Journal of the American Academy of Arts and Sciences*, **Fall**, 61–81.

Jones, R. R. and Southwood, T. R. E. (eds) (1987). *Radiation and Health*. Wiley, London.

King, J. L. (1989). Environmental regulation in the USA and the UK compared. Part 2: UK proposals for change. *Environmental Law*, **3**(4), 5.

Kramer, L. (1989). The open society, its lawyers and its environment. *Journal of Environmental Law*, **1**, 1–9.

Lashof, D. A. and Ahuja, D. R. (1990). Relative contributions of greenhouse gas emissions to global warming. *Nature*, **344**, 529–31.

Lee, T. R. (1989). Social attitudes and radioactive waste management. In *Disposal of radioactive and other hazardous wastes*. Proceedings of an international workshop on Principles for the Disposal of Radioactive and other Hazardous wastes, Stockholm, June 1988. (ed. R. Boge *et al.*) (Ds 1989:20). National Institute of Radiation Protection, Ministry of Environment and Energy, Stockholm.

Leggett, J. (ed.) (1990). *Global warming: the Greenpeace report*. Oxford University Press.

Linet, M. S. (1985). *The leukemias: epidemiologic aspects*. Oxford University Press, New York.

Mason, B. J. (1989). The greenhouse effect. *Contemporary Physics*, **30**, 417–32.

Mason, B. J. (ed.) (1990). *The surface waters acidification programme*. Cambridge University Press.

Morgan, A., Harrison, J. D., and Stather, J. W. (1990). Doses to the human fetus from plutonium intakes during pregnancy. *NRPB Radiological Protection Bulletin*, **114**, 10–14.

Murozumi, M., Chow, T. J., and Patterson, C. C. (1969). Chemical concentrations of pollutant lead aerosols, terrestrial dusts and sea salts in Greenland and Antarctic snow strata. *Geochimica et Cosmochimica Acta*, **33**, 1247–94.

Myers, N. (1990). Tropical forests. In *Global warming: the Greenpeace report*. Jeremy Leggett (ed.), pp.372–98. Oxford University Press.

O'Riordan, M. (1990). Human exposure to radon in homes. Recommendation of the practical application of the Board statement. *Documents of the NRPB*, **1**(1), 17–32.

Porter, G. (1989). Anniversary address by the President. *Royal Society News*, **5**(6) Supplement, i–vi.

Roberts, L. E. J., Liss, P. S., and Saunders, P. A. H. (1990). *Power generation and the environment*. Oxford University Press.

Royal College of Radiologists and National Radiological Protection Board (1990). Patient dose reduction in diagnostic radiology. *Documents of the NRPB*, **1**(3), 1–45.

Royal Commission on Environmental Pollution (1976a). Air pollution control: an integrated approach. *Fifth Report, Cmnd 6371*. HMSO, London.

Royal Commission on Environmental Pollution (1976b). Nuclear power and the environment. *Sixth Report, Cmnd. 6618*. HMSO, London.

Royal Commission on Environmental Pollution (1979). Agriculture and pollution. *Seventh Report, Cmnd. 7644*. HMSO, London.

Royal Commission on Environmental Pollution (1983). Lead in the environment. *Ninth Report, Cmnd. 8852*. HMSO, London.

Royal Commission on Environmental Pollution (1984). Tackling pollution—experience and prospects. *Tenth Report, Cmnd. 9149*. HMSO, London.

Royal Commission on Environmental Pollution (1985). Managing waste: the duty of care. *Eleventh Report, Cmnd. 9675*. HMSO, London.

Royal Commission on Environmental Pollution (1988). Best practicable environmental option. *Twelfth Report, Cm. 310*, HMSO, London.

Royal Society (1983). *Risk assessment*. Report of the Royal Society study group. The Royal Society, London.

Sansavini, S. (1989). Conventional or alternative agriculture? The 'integrated' and 'organic' approaches. *Alma Mater Studiorum: Universita degli Studi di Bologna*, **II**(2), 156–67.

Sears, J. (1988). Regional and seasonal variation in lead poisoning in the Mute swan *Cygnus olor* in relation to the distribution of lead and

lead weights in the Thames area, England. *Biological Conservation*, **46**, 115–34.
Sears, J. (1989). A review of lead poisoning among river Thames Mute swan *Cygnus olor* population. *Wildfowl*, **40**, 151–2.
Southwood, T. R. E. (1972). The environmental complaint—its cause, prognosis and treatment. *Journal of the Institute of Biology*, **19**(2), 85–93.
Southwood, T. R. E. (1975). Environmental problems—today and tomorrow. *Pharmaceutical Journal*, **214**, 525–6.
Southwood, T. R. E. (1980). Seventh Bawden Lecture—Pesticide usage, prodigal or precise. *Proceedings of 1979 British Crop Protection Conference. Pests and diseases*, **3**, 603–20.
Southwood, T. R. E. (1985*a*). Risk through environmental change. In *Risk, man-made hazards to man* (ed. M.C. Cooper), pp.126–38. Oxford University Press.
Southwood, T. R. E. (1985*b*). The roles of the proof and concern in the work of the Royal Commission on Environmental Pollution. *Marine Pollution Bulletin*, **16**, 346–50.
Southwood, T. R. E. (1990*a*). Risk in the natural world and human society. *Science and Public Affairs*, **5**, 85–99.
Southwood, T. R. E. (1990*b*). Surface Waters Acidification Programme: Management Group final report. *Science and Public Affairs*, **5**, 75–85.
Tarnavskii, A. (1990). Environmental legislation and the struggle of the Soviet community with the 'Flow reversal' project. *Journal of Environmental Law*, **2**, 153–9.
Tolba, M. K. (1989). The road to Montreal—and beyond. In *A modern approach to the protection of the environment: study week, November 2–7, 1987* (ed. G.B. Marini-Bettolo), pp.415–33. Pontificia Academia Scientiarum, Vatican City.
Winter, G. (1988). *Business and the environment*. McGraw-Hill, Hamburg.
World Bank (1990). *World development report 1990*. Published for the World Bank by Oxford University Press, New York.
World Resources Institute, International Institute for Environment and Development and United Nations Environment Programme (1988). *World resources 1988–89*. Basic Books, New York.
Wynne, B. (1991). After Chernobyl: science made too simple. *New Scientist*, **129**(1753), 44–6.

2
The environment: a political view
Michael Heseltine

The Rt. Hon. Michael Heseltine, MP, was elected to Parliament as Conservative member in 1966 and has held a seat ever since. He took his degree in Philosophy, Politics, and Economics at Oxford and entered politics from a background in business and publishing. In 1970 he became Parliamentary Under-Secretary of State at the Department of the Environment and following the defeat of the Conservative Party in the general election of October 1974 held the positions of Opposition Spokesman on Industry (until 1976) and the Environment (1976–1979). When the Conservatives under Mrs Thatcher were returned to office in 1979, Michael Heseltine returned to the Department of the Environment as Secretary of State; in 1983 he was appointed Secretary of State for Defence. He resigned that office over the 'Westland affair' in 1986 and returned to the back benches.

Both in and out of office, Michael Heseltine has concerned himself energetically with environmental issues. As a Minister, he drew up the Government's plans for reviving Britain's inner cities: as a local MP, he campaigned successfully against the construction of a satellite town in the countryside south of Oxford.

When he came to Oxford on 25 October 1990, to deliver his Linacre Lecture, Michael Heseltine brought with him—through a misunderstanding for which he generously accepted full responsibility—the wrong text: instead of lecturing about the politics of the environment, he addressed a rather bemused audience on the subject of Britain's relationship with the European Community. Only two weeks later, and partly because of divisions within the Conservative Party about the future of that relationship, Michael Heseltine was challenging Mrs Thatcher for the leadership of the Conservative Party. In the second round of the contest, following Mrs Thatcher's resignation as Leader, he conceded victory to John Major who, as Britain's new Prime Minister, appointed Michael Heseltine to his old post of Secretary of State for the Environment. From this, the lecture which follows—the one which Michael Heseltine should have delivered in Oxford on 25 October 1990—gains added authority.

Let me take as my theme the politics of the environment. This is a subject less often addressed than might be thought. There has, of course, been a considerable amount of discussion about the environment and its problems in recent years. However, too often in the past a single environmental issue received a great deal of attention, occasionally prompting action, but was treated in isolation. When the initiating incident or disaster had been dealt with, the environment disappeared again from political view. The killer smog of the 1950s, which led the then Conservative Government to introduce the Clean Air Acts, was a case in point. So too was the legislation to control the disposal of toxic wastes, prompted by the discovery that children were playing on local authority tip-sites on which drums containing cyanide were to be found.

This was not true only in Britain. In Italy, the explosion of a chemical plant near Seveso in 1976 contaminated a large area of land and led—six years later—to the introduction of European legislation on safety controls to prevent chemical accidents and measures to limit the consequences of any which might occur for both people and the environment. German environmental concern dates very precisely from the period in the late 1970s when it was discovered that acid rain was killing large areas of forest.

From time to time individual political leaders took an interest in some aspect of the environment, but it rarely made it onto the main agenda for governments, and did not stay long when it did. Nowhere could you have found a coherent response from government that even tried to deal with the full range of environmental issues in an integrated way. Until very recently there was, in effect, no politics of the environment.

I think there is now emerging in Britain, as elsewhere, a broad consensus that the environment is one of the greatest challenges facing us as we stand on the threshold of a new millenium. We have passed the point when it is sufficient simply to refine our understanding of our ever-more pressing problems. We must now find real and lasting solutions. This will require us to bring together sophisticated science and the strengths of our democratic tradition: firm and decisive government and the vigour, efficiency, and innovative power of free, open markets; and the enthusiasm and optimism of the young with the experience and judgement of age.

At the heart of the environmental challenge on a great range of issues—whether they are global, national, or local—is not so much the identification of solutions as the finding of ways to implement them. It is now widely accepted, not the least by this Conservative Government, that we must change the way we do many things in order to put our economy on the path towards sustainable development.

It was the report of the World Commission on the Environment and Development, *Our Common Future* (1987), that first brought the concept of sustainable development to public attention. The Ministerial Conference in Bergen in May 1991 gathered together Ministers from the 35 nations of the UN's Economic Commission for Europe—effectively the bulk of the world's industrialized nations—to discuss ways of bringing about sustainable development. The great global United Nations Conference on the Environment and Development which will take place in Brazil in 1992, on the twentieth anniversary of the Stockholm Conference, will bring together the Heads of State or Government from over 160 nations for the first-ever global summit. Its theme will be sustainable development.

As HRH Prince Charles said in his recent television film on the environment, reaching sustainable development is a task that must be accomplished by the present generation. If we fail, then the prospects for our children will be less bright than those we inherited ourselves. That would be no legacy to leave. Achieving sustainable development will require the greatest mobilization of people, capital, resources, and technology that has ever occurred in peacetime—and, if we are to bring it about, we will have to ensure that we maintain that peace. Bringing about this mobilization is the task of politics.

It is timely, therefore, that we have recently begun to see for the first time the emergence of a politics of the environment. The issue is now at the top of the political agenda, not just in

Britain or Europe, but increasingly in all parts of the world. We have seen the environment become an important part of the agenda for the meetings of the leaders of the seven leading industrialized nations, the G7 meetings. It dominated the Commonwealth Conference in Kuala Lumpur in 1989. A whole session of the Conference on Security and Co-operation in Europe 1990 was devoted to environmental matters. There has been a series of major international conferences on climate change, the ozone layer, and many other topics.

In many of these Britain has played a leading part. Indeed, two of the most successful conferences on the ozone layer recently took place in London and between them led to the most binding global agreements yet to protect the environment. What we have seen in a very short space of time is the emergence of a 'green' geopolitics. The environment has joined national security and management of the economy as one of the very few issues on the permanent agenda of national leaders.

These moves by leaders on the world stage reflect and reinforce a significant shift in public mood. Public concern about the environment has long been at a high level. In 1980 some 70 per cent of people in Britain typically said they were 'concerned' or 'very concerned' about the state of the environment. More recently that figure has moved up towards 90 per cent. The same pattern is found across Europe and elsewhere. Although there is a relationship between environmental concern and affluence—more people tend to be concerned in wealthier countries—concern is not affected by downturns in the economy. Thus, it seems, once people become concerned about the environment, they remain so.

More significantly, however, evidence is emerging that many people wish to turn this general preference for a better quality environment into real choices in real market places, where they are available. *The Green consumer guide* (Elkington and Hailes 1988) leapt immediately to the top of the best-seller list. A 1989 survey revealed that about 18 million Britons, some 42 per cent of the adult population, claim to have chosen a purchase on the basis that the goods they were buying were more environmentally benign (*The Economist*, 2 Sept 1989; p. 6).

Of even greater interest is the war that has broken out between the major retailers to present themselves as 'your greener grocer'. It is not just the naïve or over-optimistic who believe in Green consumerism. Some very hard-hearted and unsentimental managers of major retail chains have committed considerable resources to back their judgement that there is a 'green' consumer alive and well and living in Britain. You cannot enter a supermarket these days without being bombarded with 'green' advice and offered a huge range of 'greener' products. It is a rather interesting tribute to the responsiveness and adaptability of the capitalist system that the supermarkets, once the symbol of the excesses of the throw-away society, should turn up at the leading edge of the 'green' revolution.

These changes feed off each other. The higher profile given to the environment by national leaders reinforces and legitimizes already high levels of public concern. Greater public concern encourages politicians to pay more attention to the environment. This has taken place against the backdrop of even greater political change in East–West relations. The swift rise of the democratic tide in eastern Europe has created a new spirit of optimism and hope that even the most apparently intractable of problems can eventually be solved. This is most welcome; but it has also created new instabilities in the region, against which we must be careful to remain on guard.

The collapse of the Iron Curtain has also transformed the politics of the environment. We have discovered that, whatever our own management of the environment, the legacy of the past was far, far worse in the communist countries. Let me give but one example. The new Czechoslovakian Government published a report on the state of its environment in May 1990 (Moldán 1990). It found 72 per cent of its amphibians were endangered, 62 per cent of its birds, 65 per cent of its mammalian species, and about 77 per cent of its reptiles. It found that air and water quality standards were routinely broken, in one case by a staggering 3000 times. It found 70 per cent of its trees to have been damaged by air pollution.

It is no accident that the environment was one of the broadest avenues down which the march to democracy took place in eastern Europe. Naturally the environment is now high on the agenda for the newly-elected political leaders.

The lesson I draw from eastern Europe is that freedom and a good environment are as inextricably bound up with each other as wealth-creation and a good environment are. It cannot be accidental that the areas where the worst environmental degradation in the world is to be found are those where freedom is least and poverty greatest. Reinforcing the democratic impulse in eastern Europe is as important for the quality of our environment as it is for our national security.

What is now clear is that it is time to act as well as argue. The arguments will, of course, have to go on. In few fields of public policy are the uncertainties so great or the penalties—either for doing too little, too late, or for doing too much, too soon—so large. Public expectations of action from governments and others, including industry and farmers, are now very high. Failure to respond to these expectations is likely to be punished by disatisfied voters.

It is clear that environmental problems cannot be solved by countries acting alone. No problem is more intrinsically international. It is perhaps with the environment—literally our common home—that the need for an expanded role for the Council of Europe is most evident. It will be on parliamentarians everywhere that the principal burden of putting our societies on track for sustainable development will fall. It is they to whom, in democracies, governments are most directly accountable. Why not, as an experiment, convene a council of parliamentarians—drawn from the elected parliaments of all Europe—to discuss the common environmental problems of our common home?

As Europeans, we in Britain are well placed to appreciate the fundamentally international dynamic of the politics of the environment. The creation of a common policy on the environment within the European Community is one of its greatest, and least recognized, achievements. The European Community is possibly unique in its ability to draft and enforce supranational law. Nowhere else in the world have nations agreed to share an element of national sovereignty for the common good. Nowhere else is there a comparable set of binding policies to protect the environment across national boundaries. The European Community serves as a laboratory in which to develop the widest range of political tools that will be needed for planetary management. Britain must therefore play a vigorous and positive role in the development of Community environment policy.

We have already made great strides. The Government's White Paper, *This common inheritance*, published in the autumn of 1990, was the first fully comprehensive statement of environmental policy published by a British government and, indeed, one of the first by any government. Possibly its most important achievement was to set in place a machinery of government that will allow for far more integration between our environmental and other policies—an essential first step towards sustainable development.

Even in advance of the White Paper discussions, we in this country have not failed to act. But we have perhaps not communicated our achievements as well as we should. With the one exception of sulphur dioxide emissions from our power stations, our record is as good or better than that of the other member states within the Community. In nature conservation and planning policy we are the leaders. We have one of the lowest rates of prosecution for breaches of Community environment law and, when found in error by the European Court, unlike some, we correct our mistakes.

It is not just the Conservative Government that has responded creatively to the environmental challenge. The 'green' consumer concept was born in the United Kingdom and has been successfully exported to many other countries. The British environmental organizations—with their nearly 5 million members and annual expenditure in excess of £200 million—are amongst the most sophisticated and effective voluntary bodies in the world.

In a whole range of environmental technologies and services

we are recognized world leaders. Britain's environmental scientists discovered the hole in the ozone layer over Antarctica in the first place and the Hadley Centre carries out key research in the international effort to understand global warming and climate change. With such powerful, alert, and organized constituencies in this country promoting a better quality environment, it would be a very foolish government that failed to respond to, and act upon, their concerns.

The Environment Protection Act (1990) has radically overhauled British pollution control systems. Its most important provision is the establishment of Her Majesty's Pollution Inspectorate and the implementation of Integrated Pollution Control (IPC). IPC will mean that the three environmental media of air, water, and land are no longer controlled by separate systems, using different methodologies and approaches and sometimes simply shifting pollution from one medium to another. There will now be a single authorization system for industrial processes which have the potential to cause most harm to the environment and a single inspectorate to monitor and oversee it.

The IPC approach has two main advantages. For the environment it means that there will be increasing pressure to minimize emissions and waste-arisings at source—in other words greater pressure to prevent rather than cure environmental abuses. For industry it offers the inestimable advantage of simplifying, without weakening, the regulatory process, since there will now be only one agency, with one regulatory philosophy, to deal with the potentially most polluting industrial processes. Provided that there are adequate resources to staff and equip the Inspectorate—and the Government has made clear that it intends there will be—this will give Britain the most modern and effective system of pollution regulation in Europe and one which other countries and the Community as a whole might use as a model.

The Act also contains a number of other important provisions dealing with waste management and the release of genetically modified organisms into the environment. It also provides for the restructuring of the Nature Conservancy Council. The NCC had done an outstanding job in establishing British conservation practice as amongst the best in the world. But that is no reason for complacency. It is perfectly legitimate for the Government to seek to make it even stronger and more responsive to local and regional, as well as national, needs.

This is particularly true in the present circumstances, when the whole question of the future of the countryside is under debate, as agriculture and food production lose their dominant place as shapers of the rural economy. What is needed is strategic thinking, which will inevitably include a review of the role and structure of institutions. This the White Paper has undertaken but it is not, and was never intended to be, the last word on these matters. The debate is now open for all who wish to discuss and contribute.

The Government has set itself a high task for which it deserves much credit. No one should underestimate the very real difficulties of setting out to produce a government-wide response to environmental problems. It is intended that the White Paper should chart the Government's course on the environment throughout the 1990s. We are aware that the environment must be seen as a whole and that any first attempts will not be perfect. All of us have much to learn and we will continue to do so. It would help somewhat if other institutions—industrial, financial, scientific, voluntary, consumer, and so on—were now to set out their own holistic approach to environment policy.

Britain has now committed itself to stabilizing its emissions of carbon dioxide, the principal greenhouse gas, by the year 2005. This is a clear, and certainly achievable, target. The Government is right to argue, as it has, that progress on combating global warming can only be made if there is international agreement on achievable targets and thus that international targets should be set at realistic levels. I am sure that they are also right not to pay too much attention to calls to change the target date from 2005 to 2000. Frankly, to debate such matters is a distracting irrelevance that will not impress or reassure an anxious public. The important thing now is to secure international agreement and to press ahead with practical measures to curb emissions of greenhouse gases.

There is a feeling in some quarters that we can only reduce our carbon emissions at a cost of considerable pain and anguish to our economy. I am not sure that this is the correct analysis. Inevitably in politics, those whose toe is beneath the boot cry most loudly and cry first. But it is by no means correct to argue that, because governments toughen or raise standards, the economy suffers. Many of the steps we can take, for example, improving the efficiency with which we use energy, will actually save money and make firms more competitive.

Setting realistic targets is an essential precondition for international agreements. But the point I want to stress—because it is so often overlooked, or deliberately ignored—is that these targets are a floor to performance, *not a ceiling*. There will be no penalty for overshooting them. I would hope we in Britain will, because I want us to get a head-start in producing the energy-efficient goods that tomorrow's consumers will want to buy. Otherwise there may be a real threat to our economy from those nations which do set themselves ambitious targets and so encourage the more efficient use of energy that in turn will make their economies more competitive. In other words, governments, by setting the conditions of improved environmental performance, bring about the conditions in which their industry invests and improves. They therefore enable industry to compete, with up-to-date products, more effectively with companies in other countries whose governments will be doing precisely the same thing.

In the field of 'green' politics, as in so many others, the capitalist reliance on the market place will prevail. And, as can be seen from the rise of the 'green' consumer, it really does work. For here the people actually make the choice. The consumer is king. Of course, the choice must be an educated choice. It must also be an available choice. Often it will be regulated. It must be encouraged by clear standards, labelling and, where appropriate, tax incentives.

There will be some conflict between industrial pressure

groups. New industries will seek higher standards, confident that they will be effective at exploiting the new 'green' consumption patterns that are thus created. There will be others, however, who argue for the cheapest industrial processes, who will resist new regulations and requirements, and who insist upon sticking to familiar ways. But, if industries are to maintain the export markets upon which their products and their wealth depend, they cannot afford to remain half-hearted about environmental standards. The choice is whether the Americans and Japanese set the world's industrial standards alone or whether Europe is going to get in on the act as well.

Some will complain that environmental considerations impose burdens on industry. But the environmental entrepreneurs upon whom 'green' growth depends will see higher standards as helping to expand industry's markets and improve its productivity. And they will want their governments to set the standards that will create a leading edge in the world market-place. Community and international law will enforce these standards and those countries which have had lax regimes, and whose governments have been slow to introduce stiffer environmental regulations, will find their industries less well-placed.

I set out these conflicts merely to indicate the scale of the political task that faces us. It is only fair to remind ourselves that there is also a cost involved in doing nothing. But if we are committed to preserving the earth's ecological balance, we must make clear the necessity of sharing national sovereignties. The winds and the water pay no heed to national frontiers and in the context of environmental pollution, neither should we. If we are committed to higher standards, it will be necessary to make it plain that change is inevitable. Those who argue against the cost or inconvenience of change must understand that without it higher standards cannot be achieved. Let there be no doubt: higher standards are what the public wants. They will not always like the bills but they will continue to press politicians for the adoption of the policies that cause them. We must also make clear that higher standards bring wealth-creating opportunities, as well as environmental enhancements. The business of saving the world is good business. The economic renaissance on reclaimed land in many of our inner cities is just one example of the prosperity that springs up as a result of policies geared to environmental improvement.

Politicians must bring together two forces: the natural acquisitiveness of the capitalist system which will seek opportunity and profit where it can find it, and the public concern that no profits are earned at the environment's expense. These two forces, far from being incompatible, are closely linked as the emergence of the 'green' consumer shows. No one should underestimate the motive force for change from an educated public able to use effective market mechanisms.

Those who understand the working of successful capitalist economies—and which successful economies are not capitalist?—know that to invest you must earn and that to invest more you must earn more. The predictions are that we could double the population of the globe over the next 40 years; there isn't any way we can maintain environmental quality in those sorts of circumstances without very considerable and continuous investment. In Britain it is the very success of our economic turn-around that allows us the time and resources to enhance the quality of our environment. Without growth there would be little to invest and even less to distribute.

It is now becoming increasingly clear that, in the real world, the faster the economy grows the faster old, inefficient, polluting, resource-greedy plant and products can be replaced by newer, 'greener', more energy-efficient, less resource-intensive equivalents. Of course, this cannot be the old-fashioned growth at any price. Often that was not real growth at all. The dynamics of economic growth can and must be harnessed to the purpose of sustaining our environment. The task for the democratic politician is, therefore, to build on the profound changes in public attitudes that have occurred as more and more of the basic needs of western populations have been met. Higher levels of material satisfaction impose a greater sense of obligation, so that we use our wealth in a more enlightened way. Meeting this challenge requires the mobilization of all the resources of our society—the business community, the voluntary sector, the education system, local and national government, and the scientific community.

Individuals with high levels of disposable income and, increasingly, high levels of disposable capital, will be looking for products and services capable of matching their environmental concern. Businessmen will look to government to enforce higher environmental standards at home to enable them to meet more stringent standards abroad. The concerns of the 'green' consumer and the interests of the environmental entrepreneur here come together.

But we must not allow this growing public pressure to eclipse our sense of perspective. Finding solutions to our environmental problems in no way absolves us from other responsibilities. Our task is to solve an additional problem for humanity, not to ignore the all too familiar, real and very pressing problems—drought, famine, disease—which already beset us.

The quality of life, in its many dimensions, will be a key issue in the decade ahead. Meeting the environmental challenge is central to delivering a better quality of life. The White Paper is our principal tool for staking out the Conservative position on the environment. Over the past 11 years we have done much to give the people of Britain real freedom of choice. Now we must show that we are equally determined to give them a country—and a planet—worth choosing.

REFERENCES

Department for the Environment (1990). *This common inheritance: Britain's environmental strategy*. White Paper 0 10 112002 8.

Elkington, J. and Hailes, J. (1988). *The Green consumer guide*. Gollancz, London.

Moldán, B. (1990). *Environment in the Czech Republic*. Akademia, Prague.

World Commission on Environment and Development (1987). *Our common future*. Oxford University Press.

3
The greenhouse effect and global warming
John Mason

Sir John Mason, FRS, is Britain's leading meteorological scientist and the country's foremost expert on the greenhouse effect and global warming. After holding a Lectureship in Meteorology and then the Professorship of Cloud Physics at Imperial College London, Sir John was appointed Director-General of the Meteorological Office in 1965. He held that post for 18 years and presided over the computerization of Britain's weather forecasting. Since 1983, he has directed the Royal Society's Programme on the Acidification of Surface Waters. In 1986 Sir John was elected President of the University of Manchester Institute of Science and Technology (UMIST). He has published two seminal books on the physics of clouds and is the author of numerous papers on physics and meteorology.

INTRODUCTION

The greenhouse gases, especially water vapour and carbon dioxide, play a crucial role in regulating the temperature of the earth and its atmosphere. In the absence of these gases, the average surface temperature would be $-19°C$ instead of the present value of $+15°C$, and the earth would be a frozen, lifeless planet. There is now concern that atmospheric temperatures will rise further, due to the steadily increasing concentration of carbon dioxide resulting largely from the burning of fossil fuels. The concentration is now 27 per cent higher than that which prevailed before the industrial revolution, and is increasing at 0.5 per cent per annum. It is expected to reach double the 1860 value during the second half of the next century. Recently we have become aware that other strongly absorbing gases, notably methane, nitrous oxide, ozone, and chlorofluorocarbons (CFCs) are adding to the greenhouse warming and may, by the middle of the next century, contribute about half as much again as the increase in carbon dioxide.

Higher temperatures will be accompanied by changes in other climate parameters such as precipitation, cloudiness, soil moisture, snow cover and may, eventually, result in significant expansion of the oceans and melting of the mountain glaciers, and hence lead to a rise in sea level.

Although the climate changes and their economic and social effects are likely to vary seasonally, and geographically from region to region, and even from country to country, the overall problem of man-induced (anthropogenic) climatic change is a global one, whose thorough study is beyond the resources of any one country. It follows that national research programmes should be planned largely as contributions to international projects, such as the World Climate Research Programme and the World Ocean Circulation Experiment (described by Mason, 1987).

THE GLOBAL BUDGET OF CARBON DIOXIDE

In order to estimate more accurately the future concentration of atmospheric carbon dioxide, it will be necessary to study and understand in more detail the complete carbon cycle. The problem is that partition of the added man-made carbon dioxide between the atmosphere, oceans, and biosphere, involves difficult estimates of small differences between large two-way fluxes between enormous reservoirs, as illustrated in Fig. 3.1.

The total atmospheric reservoir of CO_2 is equivalent to 743 GtC* which is much smaller than 1760 GtC for the terrestrial biosphere, of which about 560 Gt is stored in trees and plants, and is tiny compared with the 39 000 Gt in the oceans. The atmospheric concentration is therefore susceptible to rather small changes in the fluxes between these reservoirs. The current rate

* GtC = gigatonnes of carbon. 1 Gt = 10^9 = 1 billion tonnes.

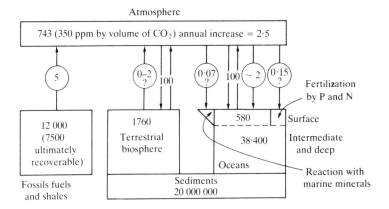

Fig. 3.1 Global carbon reservoirs and present natural and anthropogenic fluxes between reservoirs (reservoir size in GtC). Fluxes between reservoirs in GtC per year. Anthropogenic fluxes are circled. (After Liss and Crane 1983, with permission from the publisher.)

of emission of CO_2 from the burning of fossil fuels is 5.4 GtC/year, whilst the net emissions due to deforestation and changes in land use are estimated at 1.6 GtC/year. These are small compared with the fluxes exchanged between the atmosphere and the earth's surface, which exceed 200 GtC/year. The atmosphere retains about 3.4 Gt (almost 50 per cent of the emissions) leaving 3 Gt/year to be taken up by the oceans. The net fixation of CO_2 by photosynthesis, largely by phytoplankton growing in the top 100 m or so, is about 100 GtC/year, about the same as for the terrestrial biosphere. Most of this is released by respiration and returns to the atmosphere, but some is dissolved in the ocean and some is converted into inorganic carbon in the skeletons and faeces of zooplankton and falls to the ocean floor.

Models of the ocean uptake suggest that it can accept about 1.8 GtC/year, so that there is an apparent imbalance of about 1.6 GtC/year. This is a measure of the uncertainty in current understanding of the global budget of atmospheric carbon dioxide. Either there are some as yet unidentified mechanisms for removing CO_2 from the atmosphere, or the amount of CO_2 released by tropical deforestation has been greatly overestimated, and/or our quantitative knowledge of the known mechanisms is unsatisfactory. Nevertheless, relatively minor adjustments in the world ocean circulation and chemistry are likely to affect significantly the amount of CO_2 added each year to the atmosphere, even if emissions are stabilized. In particular, ocean warming is likely to decrease the net uptake of CO_2 by sea-water. Until this problem is resolved, predictions of the proportion of future emissions of CO_2 retained in the atmosphere will be subject to considerable uncertainty. However, an even larger uncertainty lies in the future global rate of increase of combustion of fossil fuels and wood, for which estimates range from less than 2 per cent per annum to double this figure.

CARBON DIOXIDE IN THE ATMOSPHERE

Analysis of air bubbles trapped in the deep interiors of glaciers reveals that the atmospheric concentration of CO_2 in the last ice age was about 210 ppmv.* The value at the beginning of the industrial revolution is estimated at 275 ppm and is estimated to have increased by 15 per cent over the following 100 years to reach 316 ppm in 1960. Since accurate and continuous measurements were started in 1958, the concentration has risen at an ever-increasing rate (see Fig. 3.2), which is currently very nearly 0.5 per cent per annum. The present concentration of 354 ppm is 27 per cent above the 1860 value. If the concentration were to continue to increase at the present rate (0.5 per cent per annum), it would double its pre-industrial value by 2080, and double its present value by 2130 AD. However, it is likely that the increase will continue to accelerate, particularly if the world's population continues to increase at the current rate, and it may reach double the present value, that is 700 ppm, in 80–130 years, depending on the future rate of burning fossil fuels

* ppmv = parts per million (in the atmosphere as measured by) volume.

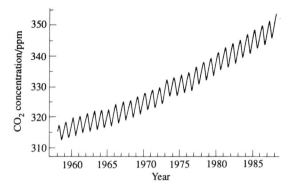

Fig. 3.2 The increase in atmospheric concentrations of CO_2 since 1958, as measured at Mauna Loa, Hawaii. (Reproduced with permission from *Contemporary Physics*.)

and wood and the extent to which CO_2 is taken up and stored in the oceans and by trees and vegetation through photosynthesis.

THE EFFECT OF CO_2 AND WATER VAPOUR ON THE RADIATION BUDGET OF THE ATMOSPHERE

The heat budget of the atmosphere and earth is shown diagrammatically in Fig. 3.3 Of the short-wave solar radiation incident on the top of the atmosphere, a global annual average of 340 W/m^2,* about 17 per cent is reflected back to space by clouds, 8 per cent is back-scattered by the air, and 6 per cent reflected by the earth's surface, to give a planetary albedo of 31 per cent.† About 19 per cent is absorbed by water vapour, ozone, and dust and about 4 per cent by clouds as the radiation passes through the atmosphere, so only 46 per cent is absorbed at the surface.

---

* W/m^2 = Watts per square metre. 340 W/m^2 is the mean energy of the solar radiation falling on unit area of the atmosphere's surface during one second.
† Albedo is the total fraction of radiation striking the earth which is reflected back into space.

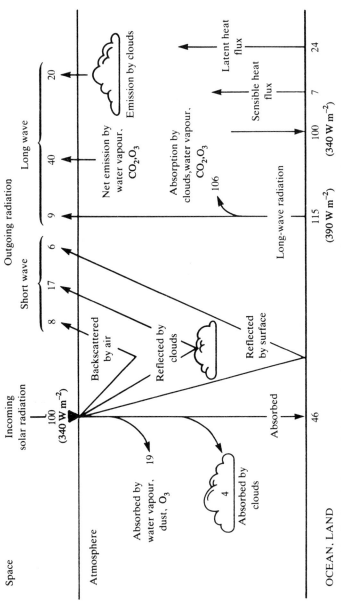

Fig. 3.3 Radiation balance of the earth: the numbers represent fluxes expressed as percentages of the global average incoming solar flux at the top of the atmosphere (100 m units = 340 W/m²).

This is transferred to the atmosphere as infra-red radiation (a net 15 units, where 100 units represents the incoming solar radiation), as sensible heat (seven units), leaving 24 units to evaporate water which later condenses to form clouds and a global annual average rainfall of 104 cm.

Considering now the balance of the long-wave radiation on the right-hand side of Fig. 3.3, of the 115 units (390 W/m^2) emitted by the earth's surface, only nine are transmitted through the atmospheric window to space, the other 106 being absorbed by the atmosphere, mainly by water vapour, carbon dioxide, and ozone. This absorption of the up-welling long-wave radiation, plus that from the incoming solar radiation (19 units), plus the sensible heat flux (seven units), total 132 units to which must be added a net contribution of $(24 + 4 - 20 = 8)$ units from clouds. Of this total heat input of 140 units, the atmosphere emits 40 to space and 100 (340 W/m^2) to the surface, the net long-wave radiative flux from the surface to the atmosphere being only 15 units. The net absorption of infra-red radiation by the greenhouse gases is the difference between the 115 units of outgoing radiation from the earth's surface and the 69 units emitted at the top of the atmosphere, that is 46 units or 154 W/m^2.

In the absence of absorbing greenhouse gases (mainly water vapour and CO_2), the equilibrium black-body surface temperature T_e of the planet, assuming it to have an albedo of $\alpha = 0.31$, is given by the equation known as Stefan's Law:

$$4\pi a^2 \cdot \sigma T_e^4 = \pi a^2 S(1 - \alpha)$$

or

$$T_e = \left[\frac{S(1 - \alpha)}{4\sigma}\right]^{1/4} = 254 \text{K}$$

where 'σ' is Stefan's constant, 'S' the solar constant and 'a' the radius of the planet.*

The intensity of the emitted radiation would be only 236 W/m^2 compared with the actual value of 390 W/m^2 which again

---

* Stefan's constant = 5.6697 x 10^{-8} W/m^2/K^4. This is the constant of proportionality in Stefan's Law.

implies that the combined contribution of the greenhouse gases is 154 W/m^2. About 100 W/m^2 are calculated to come from water vapour and about 50 W/m^2 from carbon dioxide.

Both water vapour and carbon dioxide absorb infra-red over a range of wavelengths, and the relatively high concentrations of these two gases ensure that many of the spectral lines are saturated and that any increase in absorption from an increase in their concentration is limited to the wings of the absorption lines. Thus, while the present concentration of CO_2 (354 ppmv) produces a downward atmospheric flux of 50 W/m^2, a near doubling to 600 ppmv would increase the flux by only 4 W/m^2 and raise the surface temperature by only 1.2 K in the absence of feedback effects due to water vapour, clouds, ice, etc.* The quantitative impact of the various feedback mechanisms in the global climate system can be estimated only from the results of large, complex models described below. However, a simple calculation for a climate system in thermal equilibrium indicates that the concomitant increase in water vapour would amplify the temperature rise due to CO_2 by a factor of 1.6 to reach 1.9 K. On the same basis, the enhancement of CO_2 since 1860 should have produced a warming of about 0.6 K.

THE ROLE OF THE OTHER GREENHOUSE GASES

Although water vapour and carbon dioxide are the main cause of the greenhouse effect, any gas that absorbs in the infra-red will help to reduce the loss of terrestrial radiation to outer space. However, absorption by water vapour and carbon dioxide is so strong that other gases will contribute little unless they absorb at wavelengths, mainly from 8 μm to 12 μm (the atmospheric 'window'), where absorption by CO_2 and water is weak.

The most important of the trace gases which contribute signi-

---

* K = degrees Kelvin. The Kelvin scale of temperature has the same unit size as the Celsius scale, but the zero of the scale is at absolute zero (−273,15°C)—the lowest temperature theoretically possible.

ficantly to the trapping of terrestrial radiation, despite their small concentrations, are methane (CH_4), nitrous oxide (N_2O), and the chlorofluorocarbons (CFCs), especially CCl_3F (F11) and CCl_2F_2 (F12). Methane concentrations are only about 0.5 per cent as high as those of CO_2, but methane is 21 times as effective, molecule for molecule, than CO_2 as an absorber of terrestrial radiation. Similarly the CFCs, whilst present at much lower atmospheric concentrations even than methane, are some 14 000 times more effective than CO_2 at absorbing radiation.

Methane, currently at 1750 ppbv,* is increasing at about 1 per cent annually, and is expected to double in about 70 years. Various sources of methane have been identified but not quantified. These include emissions from ruminants, rice paddies, waste disposal sites, and oil recovery operations. Doubling methane would have about 15 per cent of the warming effect of doubling carbon dioxide.

Nitrous oxide, currently at 300 ppbv, is increasing at 0.25 per cent per annum for largely unknown reasons. It is formed primarily as a product of bacterial denitrification but also in combustion processes. It is likely to increase by about 20 per cent by 2060, contributing in the meantime about 4 per cent to the total greenhouse warming.

Table 3.1 shows the present concentrations of CFCs in the atmosphere and their contributions to global warming to be about 0.3 W/m^2, or 12 per cent. The trace gases combined now

Table 3.1 Global warming by greenhouse gases in 1990 relative to 1765

	CO_2	CH_4	N_2O	CFC_{11}	CFC_{12}	$HCFC_{22}$	
Concentrations (1765)	279	790	285	0	0	0	
Concentrations (1990)	354	1720	310	280	480	320	
	ppm	ppb	ppb	ppt	ppt	ppt	
Increased heat flux	1.5	0.42	0.10	0.06	0.14	0.08	$\Sigma = 2.30 \text{ W/m}^2$
% contributions	66	18	4		12		

* ppbv = parts per billion (in the atmosphere as measured by) volume. 1 billion = 1000 million.

contribute about half as much again as CO_2 to greenhouse warming. Even if CFC emissions are reduced, the concentrations may continue to rise because F11 and F12 have atmospheric lifetimes of 80 and 140 years, respectively. If the Montreal Agreement to reduce emissions to 80 per cent of 1986 levels from 1993, and to 50 per cent from 1998, is fully implemented, CFCs are likely to contribute almost 10 per cent to greenhouse warming in 2060.

A simple radiative calculation with no feedbacks would suggest that those greenhouse gases together should have produced a global warming of 0.7 K relative to 1765, but enhanced to 1.1 K by the concomitant increase in water vapour. A corresponding calculation of the present warming relative to 1900 gives 0.85 K, which is very close to the best estimate obtained from advanced models, of 0.9 K. However, this agreement is probably fortuitous as the simple calculations ignore all feedbacks except that due to water vapour.

IS GREENHOUSE WARMING APPARENT IN THE OBSERVATIONS?

The estimate that the climate should have warmed by almost 1 K since 1900 prompts one to examine the observed average global temperature record which, as shown in Fig. 3.4, indicates a rise of about 0.5 K over the last 90 years. However, it is unlikely that this can be attributed to the greenhouse effect for the following reasons:

(a) 0.3 K of the 0.5 K rise occurred between 1900 and 1940 when CO_2 was increasing at only 0.1 per cent per annum, compared with the current rate of 0.5 per cent;

(b) there was a small *fall* in temperature between 1940 and the mid-1960s (widely claimed by the media and some scientists in the 1970s to herald a new ice age);

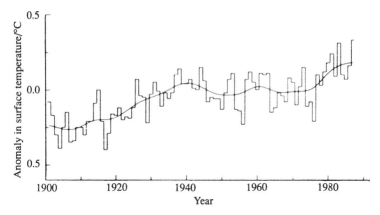

Fig. 3.4 Globally averaged anomalies in surface temperatures for 1900–87 relative to the average values for 1950–79. (Based on data from the Meteorological Office and the Climate Research Unit, University of East Anglia.)

(c) despite the rather sharp rise during the last decade in average *global* air temperatures, this has not occurred in high latitudes, nor has there been any significant decrease in the ice cover, although all the models predict greatest greenhouse warming in the Arctic.

The timing of the fluctuations in the temperature record, and the fact that greenhouse warming is likely to be delayed for some decades because of the thermal inertia of the oceans (see below), strongly suggest that these are natural climatic fluctuations. If we assume that this is indeed the case, we may use a coupled atmosphere–ocean model to estimate how long it will take for a greenhouse signal of, say, 0.5 K to be detectable. The only published account (Manabe *et al.* 1990) of a coupled atmosphere-deep ocean model, in which CO_2 is allowed to increase at 1 per cent per annum compound and so double in 70 years, predicts an average global temperature rise of 2–3K, a rise of 0.5 K occurring after 20 years.

MODEL SIMULATIONS AND PREDICTIONS OF CLIMATE CHANGE

Changes in global and regional climates due to greenhouse gases will be small, slow and difficult to detect above natural fluctuations during the next 10–20 years. It will therefore be necessary to rely heavily on model predictions of changes in temperature, rainfall, ice cover, and so on. Indeed, in the absence of any direct evidence, concern over the greenhouse effect is based almost entirely on model predictions which, unfortunately, vary so widely that they do not yet provide a sufficiently firm basis for government action.

Climate models, ranging from simple one-dimensional energy-balance models to enormously complex, three-dimensional global models requiring vast computing power, have been developed during the last 20 years, the most advanced of which are situated at three centres in the USA and the UK Meteorological Office.

Until recently, effort was concentrated on developing models (that evolved from weather prediction models) of the global atmosphere coupled to the oceans and cryosphere (sea and land ice) only through prescribing and updating surface parameters such as temperature and albedo, from observations. However, realistic predictions of long-term changes in climate, natural or man-made, must involve the atmosphere, oceans, cryosphere and, eventually, the biosphere, treated as a single, strongly coupled and interactive system. The oceans play a major stabilizing role in global climate because of their inertia and heat-storage capacity. Moreover, they transport nearly as much heat between the equator and the poles as does the atmosphere. The oceans will delay warming by the greenhouse gases because they absorb about half of the CO_2 emitted by burning fossil fuel, and also absorb and transport a good deal of the associated additional heat flux from the atmosphere.

The Meteorological Office has developed perhaps the most advanced model of the global atmosphere coupled to simple models of the ocean, and of land and sea ice, and have used these to study the effects of nearly doubling the present level of CO_2 to 600 ppmv.

The atmospheric model

The physico-mathematical models of the atmosphere are based on the physical and dynamical laws that govern the birth, growth, decay, and movement of the main weather systems that can be resolved by the model. In other words, the models must properly represent the relevant or significant scales of motion and their non-linear reactions, but smooth out all the smaller-scale motions that cannot be adequately observed or represented individually. However they must allow for the overall contribution of smaller scale motions to transport and energy conversion processes by representing their statistically averaged properties in terms of larger-scale parameters that can be measured. The parameterization of these sub-grid scale processes is one of the most difficult and uncertain features of weather and climate models, and occupies a good deal of the present research effort.

The models incorporate the principles of conservation of mass, momentum, energy, and water in all its phases, the Newtonian (Navier-Stokes) equations of motion applied to a parcel of air, the laws of thermodynamics and radiative transfer, and the equation of state of humid air. Parameters specified in advance include the size, rotation, geography, and topography of the earth, the incoming solar radiation and its diurnal and seasonal variations, the radiative and heat conductive properties of the land surface according to the nature of the soil, soil moisture, vegetation, and snow or ice cover, all of which are computed every five days.

The atmosphere is divided into 11 concentric shells (11 levels) between the surface and 20 mb* (about 30 km) with three levels in the surface boundary layer (lowest km) to allow calculation

---

* mb = millibar (unit of atmospheric pressure).

of the surface fluxes of heat, moisture, and momentum. There are also three levels in the soil to calculate the heat flux through the soil and hence the land surface temperature. The variables are calculated on a spherical grid with mesh 2½° latitude × 3¾° longitude with some 30 000 points at each level, or about 350 000 points in all.

The main physical processes represented in the model are the following.

(a) Transfer of heat by:
  (i) solar and terrestrial (infra-red) radiation, including absorption by the greenhouse gases, water vapour, carbon dioxide, ozone, and methane, scattering and absorption by clouds, reflection/absorption at the earth's surface by soil, vegetation, snow, land and sea ice, and by the oceans;
  (ii) shallow and deep convection;
  (iii) conduction at the earth's surface.

(b) The hydrological cycle:
  (i) evaporation of moisture from land and water surfaces, condensation in the atmosphere to form clouds, rain, and snow (precipitation), calculation of run-off and soil moisture;
  (ii) transport of heat, moisture, and momentum in the lowest atmospheric layers (atmospheric boundary layer) by small-scale turbulent motions;
  (iii) the frictional drag on the atmosphere exerted by mountains, the land surface, breaking gravity waves in the atmosphere, and by waves on the ocean surface.

Starting from initial values derived from observations on a particular day, the governing finite-difference equations are integrated in time-steps of 20 minutes to give new values of the following parameters at all the relevant grid points, which are then averaged to give monthly mean values over total integration times of years or decades.

The most important computed variables are:

- East–West and North–South components of the wind
- vertical motion
- air temperature and humidity
- heights of the 11 specified pressure surfaces
- short- and long-wave radiation fluxes
- cloud amount, height, and liquid-water content
- precipitation—rain/snow
- atmospheric pressure at the earth's surface
- land surface temperature
- soil moisture content
- snow cover and depth
- sea-ice cover and depth
- ice-surface temperature
- sea-surface temperature.

The atmospheric model is coupled to a simple, shallow well-mixed ocean layer only 50 m deep, which transports heat horizontally and vertically and allows computations of the fluxes of heat, moisture, and momentum between the ocean surface and the atmosphere. An energy-balance sea-ice model allows the areal cover, snow depth, and ice thickness to be calculated every five days and the albedo to decrease gradually as the ice melts and recedes. The surface temperature of the sea-ice is calculated from the heat-balance equation at every model time-step.

A 24-hour integration for the whole system involves about 10^{12} numerical operations, so that a complete annual cycle takes about ten hours on the most powerful supercomputer, the CRAY YMP.

Such models have been remarkably successful in simulating the main features of the present global climate—the distribution of temperature, rainfall, winds, and so on—and their seasonal and regional variations. They do, however, contain some systematic errors (different in different models), and will require continued development and improvement in order to provide accurate simulations/predictions of the fractionally small but, nevertheless, potentially very important changes that may result from natural or man-made perturbations.

Model simulations for the doubling of CO_2

The changes in global temperatures and precipitation to be expected from a near-doubling of carbon dioxide to 600 ppm, as predicted by the Meteorological Office model before the recent changes in the treatment of clouds described below, are shown in Figs 3.5–7. These changes, achieved when the climate system represented by the model has come into equilibrium with the enhanced atmospheric CO_2, are the highest predicted by any of the advanced models. A detailed account is given by Wilson and Mitchell (1987). Although recent modifications to the cloud parameterization scheme have produced smaller changes, these 'upper' estimates are presented here because they bring out more clearly the geographical and seasonal patterns of the changes which are not greatly altered in the newer simulations.

The global average temperature is increased by 5.2 K, accompanied by a 15 per cent increase in both global precipitation and evaporation of water from the surface. Comparable results from other advanced models are shown in Table 3.2. The enhanced radiative heating of the surface due to increases in carbon dioxide and water vapour causes increased evaporation (which restricts the temperature rise), and produces a more intense, globally-averaged hydrological cycle. Enhanced CO_2 causes increased emission of long-wave radiation from the top of the atmosphere to space and, consequently, a cooling of the stratosphere (Fig. 3.5).

The zonally-averaged (along latitude bands) warming is generally most pronounced in high latitudes near the surface in winter because of several amplifying factors (Fig. 3.6(a)). Firstly, greenhouse warming reduces the highly reflecting snow and ice cover and leads to greater absorption of heat, which accelerates the retreat of the ice. Secondly, in high latitudes in winter, there

Table 3.2 Global mean changes caused by doubling CO_2 as predicted by various models under equilibrium conditions with different cloud schemes. (All models consist of a global atmosphere with 9–11 levels and a shallow mixed-layer ocean with prescribed heat transport.)

Model	Cloud representation	Radiative properties of clouds	Temperature rise K	Precipitation increase %
UKMO (1)	Empirical-linked to relative humidity; all-water clouds	Fixed	5.2	15
UKMO (2)	Computed liquid-water and ice content	Fixed	3.2	8
UKMO (3)	Computed liquid-water and ice content	Variable-function of water and ice content	1.9	3
GFDL	Empirical-linked to relative humidity	Fixed	4.0	8
GISS	Empirical-linked to relative humidity	Fixed	4.8	13
CCC	Empirical-linked to relative humidity	Variable	3.5	4

UKMO: UK Met. Office
GFDL: Geophysical Fluid Dynamics Laboratory, Princeton
GISS: Goddard Institute of Space Studies
CCC: Canadian Climate Centre, Toronto

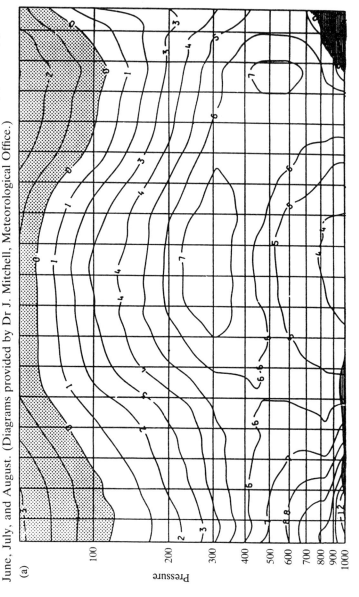

Fig. 3.5 (a) Changes in the zonally-averaged air temperatures (K) as functions of height and latitude, produced by doubling CO_2 concentrations from 320 to 640 ppm in the Meteorological Office model (Wilson and Mitchell 1987), for the months of December, January, and February combined. Areas of cooling are shaded. (b) As for (a) but for June, July, and August. (Diagrams provided by Dr J. Mitchell, Meteorological Office.)

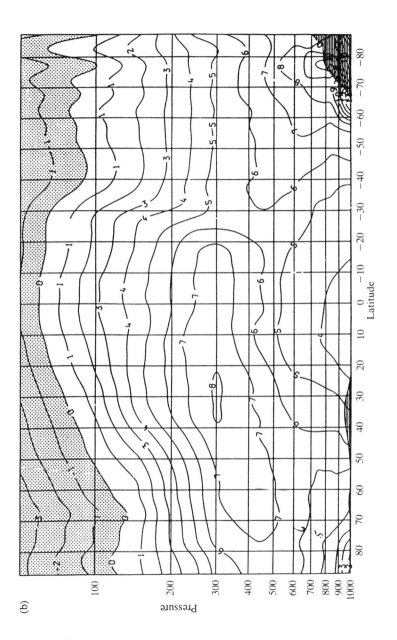

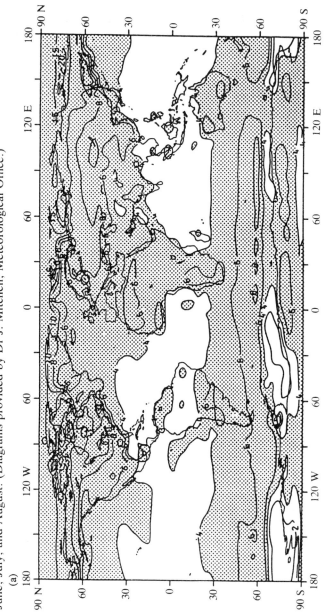

Fig. 3.6 (a) Changes in air temperature near the surface as the result of doubling CO_2 concentrations for December, January, and February combined. Areas where warming exceeds 4 K are shaded. (b) As for (a) but for June, July, and August. (Diagrams provided by Dr J. Mitchell, Meteorological Office.)

(a)

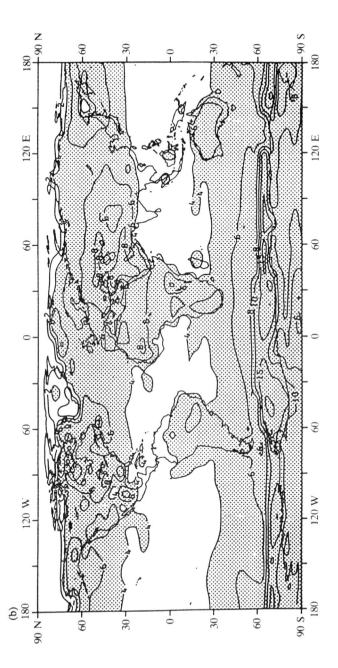

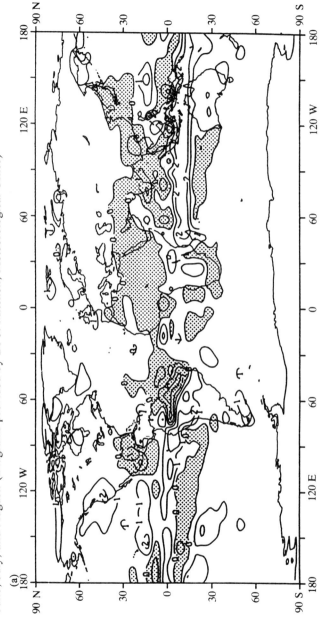

Fig. 3.7 (a) Changes in precipitation at the surface (in mm day^{-1}) as the result of doubling CO_2 concentrations for December, January, and February combined. Areas of precipitation decrease are shaded. (b) As for (a) but for June, July, and August. (Diagrams provided by Dr J. Mitchell, Meteorological Office)

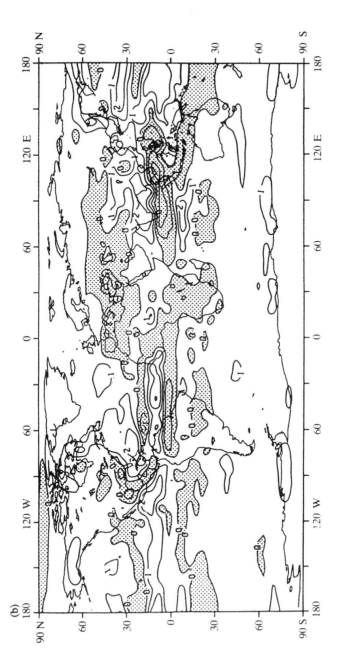

is a shallow layer of cold, dense air near the surface (an inversion) which traps the increased radiative heating, whereas in the tropics and sub-tropics this is mixed through the whole depth of the atmosphere by deep convection, to produce a smaller warming. Thirdly, the warmer, moister atmosphere transports more latent energy upwards and polewards from the tropics, releasing additional latent heat by condensation in the middle and upper troposphere on the way. Together, these positive feedback effects result in winter polar surface temperatures being increased by 12 K, compared with only 4 K in the tropics.

The simulated changes in temperature also vary considerably with longitude and season. Greenhouse warming is greatest over sea-ice in winter, smallest over sea-ice in summer (Fig 3.6(b)). In summer, sea-ice in the Arctic is maintained at constant temperature by melting. Greenhouse warming either produces more melting with no change of temperature, or melts sea-ice completely to expose an oceanic mixed layer which warms only slowly because of its large thermal inertia. The additional heat stored in the mixed layer is released in autumn and winter, delaying the onset of freezing and leading to thinner sea-ice through which heat from the ocean can diffuse more rapidly and enhance surface warming in winter. There is also considerable variation in the predicted greenhouse warming within individual continents. In regions where the soil becomes drier, evaporation may be restricted, leading to increased warming, and vice-versa.

The increase in atmospheric moisture accompanying the warming due to enhanced CO_2 would lead one to expect increased precipitation, especially in regions where the low-level winds converge to produce rising motions, notably in the extra-tropical depression belts and along the inter-tropical convergence zone. This is confirmed in Fig. 3.7, which also shows an increase in the summer monsoonal rains over South East Asia. There is also a general increase in precipitation in high latitudes, especially in winter, consistent with increased transport of moisture from low latitudes. Fig. 3.7 also shows large areas of slightly *decreased* rainfall, especially in the sub-tropics in

December–February and over Euro-Asia in the northern summer. These large-scale latitudinal and seasonal changes are broadly reproduced in the various models, but they show considerable differences in their predictions on regional and small scales. Models of higher resolution and improved physics will be required to resolve these differences. This is important because global or zonal averages are of little use in assessing the effects of greenhouse warming on agriculture, forestry, energy consumption, water supply, etc.

Sensitivity of models

It is important to note that the model results are quite sensitive to the representation of the many interacting physical processes, in particular, the simulation of clouds and their influence on the incoming solar and outgoing terrestrial radiation, to the fluxes of heat and moisture between the oceans and the atmosphere, to the positive feedback between changes in ice cover and ocean or atmosphere temperature, and changes between melting snow cover and soil moisture.

This sensitivity of the model simulations, and the differences or deficiencies of the various models in treating these physical processes, are largely responsible for the rather wide range of predicted changes of climatic parameters due to prescribed increases of carbon dioxide shown in Table 3.2.

Climate models are particularly sensitive to the parameterization of clouds, which may affect the computed equilibrium greenhouse warming by a factor of more than two. Important factors are the coverage, height, and type of cloud; their optical thickness, their reflectivity, absorptivity, and emissivity, which are determined by the concentration and size of the droplets and ice particles. Depending on these factors, clouds can generate a feedback effect that is either positive or negative. Thus a relative increase in high cloud, which has a relatively low albedo and emits less radiation, tends to warm the atmosphere, whereas a relative increase in low cloud with its higher albedo and larger emission of radiation to space, will have a cooling effect.

Overall, clouds have a net cooling effect on the planet of about 15 W/m² so that, on a simplistic view, doubling of CO_2 could be completely offset by a 20 per cent increase in the present cloud distribution, but by a smaller percentage increase if this were mainly low cloud.

The recent introduction of a more realistic representation of clouds and their radiative properties in the Meteorological Office model has led to a marked reduction in the predicted warming due to enhanced CO_2. For the model simulations described in the last section, the cloud cover was calculated from an empirical formula on the basis of the prevailing relative humidity, but the liquid-water content of the cloud and its detailed radiative properties were not computed.

The new version of the model calculates the liquid-water content of the cloud as the difference between that formed by condensation in air cooled by vertical ascent and that released as precipitation. Also, ice crystals are introduced above the $-15\,°C$ level in concentrations which increase with decreasing temperature. Furthermore, the different radiative properties of the water droplets and ice crystals are taken into account. These changes result in an atmosphere with doubled CO_2 producing an increase in low- and medium-level cloud cover at middle latitudes, and a reduction in the global average greenhouse warming from 5.2 K to only 1.9 K (Mitchell *et al.* 1989). The end result is to change the Meteorological Office predictions from being the highest to the lowest of those shown in Table 3.2.

Other important feedback effects result from the melting and retreat of the ice and snow cover induced by greenhouse warming. As the highly reflecting sea-ice retreats and is replaced by more strongly absorbing sea-water, the latter is warmed, causes further melting of ice, and so enhances the greenhouse warming. The greenhouse warming also accelerates the melting and retreat of the snow cover at middle and high latitudes in the spring. The newly exposed soil is then subject to greater warming and evaporation and the soil moisture is reduced.

Depending on the model used, simulations indicate that the

combined effect of these positive feedback mechanisms would result in warming 1.5 to 3.5 times greater in magnitude than would arise from the purely radiative effects of the greenhouse gases (including water vapour). These figures correspond to an average global warming of between 2 K and 5 K if CO_2 concentrations were to nearly double to 600 ppmv.

By the time CO_2 reaches this concentration, the other greenhouse gases will have increased these figures by at least 50 per cent, even if the Montreal Agreement is carried out.

Irrespective of the actual magnitudes of the climate changes predicted by the various models, it is important to realize that they involve only very small percentage changes in the normal radiative fluxes—much smaller than the errors either in measurements or in calculations of these fluxes. We therefore have to rely on the assumption that systematic errors will be the same in the 'perturbed' and normal (control) models, and so will cancel out when we calculate the differences between the numerical predictions of the two models. It is these differences which may be attributed to the greenhouse effect.

Furthermore, these greenhouse 'signals' are comparable in magnitude with the interannual variations that occur both in the models and the real atmosphere, so it is necessary to assess whether they are significant relative to the natural noise. Such statistical tests are made as a routine in the Meteorological Office model. The results described here and shown in Figs 3.5–3.7 are significant at the 90 per cent confidence level.

Limitations and future model development

We have already intimated that current models contain some important deficiencies which lead to uncertainty in the prediction of greenhouse effects. They should therefore be taken as only broad indications of likely changes; little weight should be given at this stage to the actual magnitudes, and even less to their significance on regional and smaller scales. The hope is that continued development of the different models will cause their predictions to converge, to narrow the range of uncertainty, and thereby to provide more reliable guidance for remedial action.

Improvements are likely to come mainly from better representation of the physical processes and greater spatial resolution. The former will require more intensive study of the processes in the real atmosphere using highly-instrumented research aircraft, radars, lasers, etc. Higher resolution, the doubling of which increases the amount of computation at least eight-fold, requires greater computing power than is currently available. One therefore has to choose, or effect a compromise, between the need for high resolution and more detailed physics leading to greater accuracy on the one hand, and the need for long model runs to study interannual and longer-term variations in climate on the other.

These longer-term changes, on time-scales of decades to centuries, will be largely determined by the oceans, not only in the surface layers, but at depth. It is therefore necessary to develop fully three-dimensional models of the global ocean circulation which are coupled to, and driven by, the winds of the atmospheric model. Such models are being developed in the USA and the Meteorological Office in the UK with encouraging success and should be ready to utilize data on sea-surface temperature, surface winds and wind stress, and circulations in the surface layers of the ocean, from the new oceanographic satellites in the early 1990s and beyond. However, as yet very few long-term integrations have been made with these fully-coupled models, which will require even more computing power, especially if it proves necessary to resolve the ocean 'weather' systems such as gyres, eddies, and fronts that are an order of magnitude smaller in linear dimension than their atmospheric counterparts.

The improvement of climate models will also depend on an adequate supply of observations from all parts of the climatic system, not least the atmosphere. These are required for input to and initialization of the models, for their validation, and for the detection and monitoring of climate changes in the climate system itself. Provision of these observations is one of the main functions of the World Climate Research Programme and the

World Ocean Circulation Experiment (described by Mason, 1987).

TIME-DEPENDENT SIMULATIONS OF GREENHOUSE WARMING—THE DELAYING EFFECTS OF THE OCEANS

Virtually all the model calculations on greenhouse warming to date have assumed that both the normal climate system (control model) and the system perturbed by enhanced levels of greenhouse gases are in equilibrium at all stages. The carbon dioxide is doubled in one step and the model climate system is allowed to come into equilibrium with the new concentration. In the real world this will never be the case because the trace gases are increasing gradually with time and the response of the total climate system depends upon a variety of physical and biogeochemical processes acting on widely different timescales. The atmosphere, together with the sea-ice, the upper layers of the ocean and the land-surface hydrology, respond quite rapidly and reach a quasi-equilibrium in the model after a few annual cycles. On the other hand, the deep ocean circulation and the land-based ice sheets respond much more slowly on time-scales of hundreds of years, and so will constantly lag behind the response of the atmosphere.

The large thermal capacity of the oceans and their ability to store and transport some of the additional heat flux from the trace gases will delay the greenhouse warming but also ensure that temperatures will continue to rise long after any reduction in emissions takes place. This is probably the main reason why there is, as yet, no convincing observational evidence of temperature rises due to the gases accumulated so far.

The delaying effect of the oceans will be determined by the net additional heat flux at the ocean surface produced by the greenhouse gases and the effective heat capacity of the ocean which is determined by the penetration of the heat below the surface. Initially the warming will involve only the well-mixed, stably-stratified, 'warm water shell' down to a mean depth of 100 metres or so and, even after 20 years, may involve only the top 500 metres, as indicated by the observed penetration of tritium ejected into the atmosphere during the series of thermonuclear explosions in the 1950s.

The actual changes in climatic parameters for a given increase in greenhouse gas concentrations are therefore likely to be smaller than those predicted by the 'equilibrium' models, and to be delayed. More realistic estimates of the magnitude and timing of the greenhouse effects will require a model which couples the atmosphere to a global deep ocean, and in which the concentrations of gases are permitted to increase gradually at current or predicted rates. Only a very few model simulations of this type have been published. The results of one experiment by Manabe *et al.* (1990) in which the CO_2 was increased at 1 per cent per annum compound to double in 70 years, are shown in Fig. 3.8, which plots the increase in surface air temperatures over the globe, averaged over one annual cycle. The globally averaged value of 2.3 K is much lower than that obtained by all other advanced models, except the latest Meteorological Office version mentioned above. The reduced warming is especially marked in the southern hemisphere, which shows little enhanced warming in the Antarctic compared with that in the Arctic. This is explained by the vertical ocean circulation in the southern oceans, which produces a deep down-welling of water around 65°S* that carries much of the additional greenhouse flux of heat from the surface to great depths (see Fig. 3.9) where it is stored for many decades. This very interesting result should be treated with caution because this is only the first experiment with a model that, like all coupled ocean–atmosphere models, has considerable difficulty in reproducing the correct fluxes of heat and moisture at the interface. However, it points to the moderating and delaying effect of the oceans on greenhouse warming of the atmosphere. According to this simulation, a temperature rise of 0.5 K would occur after 20 years.

* °S is a measure of latitude. The Equator is taken to be at 0°, the North Pole at 90°N, and the South Pole at 90°S.

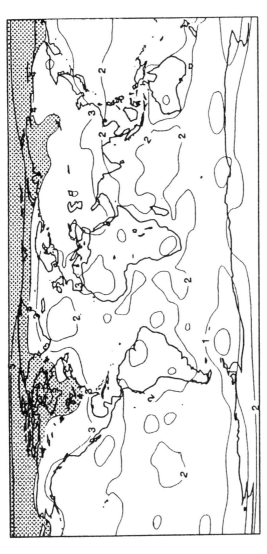

Fig. 3.8 The predicted changes in air temperature (°C) produced by CO_2 increasing at 1 per cent p.a. for 70 years according to the coupled ocean atmosphere model of Manable *et al.* (1990). (Reproduced from Houghton *et al.* 1990, with permission from Cambridge University Press).

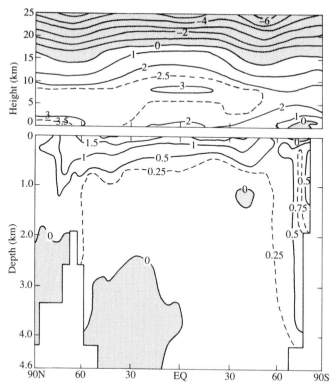

Fig. 3.9 The changes in zonal mean temperatures in both the atmosphere (above) and the oceans (below) brought about by increased CO_2 at 1 per cent p.a. for 70 years according to the coupled atmosphere–ocean model of Manabe *et al.* (1990). (*Source*: Houghton *et al.* 1990)

Just when, during the next century, the greenhouse gases will reach concentrations at which they will produce significant climate changes, will depend also on the rates of exchange of the gases between the atmosphere and the oceans, the fraction of gases retained in the atmosphere, the take-up, storage, and release of CO_2 by phytoplankton, forests, and other vegetation, all involving biological processes that will have to be incorporated into global climate models at a suitable stage. The first requirement is to obtain high-quality observational data, which is one of the objectives of the International Geosphere Biosphere Programme. The National Biogeochemical Ocean Fluxes Study, which is part of the international Joint Global Ocean Fluxes Study, and which will form a strong interface with the World Climate Research Programme, is an important first step along this road. However, by far the most important factor in determining the future levels of atmospheric CO_2 and other greenhouse gases, and hence the timing of significant climate changes, will be the future rates of emissions. Scenarios for these differ so widely that estimates of the time likely to elapse before the CO_2 reaches double the present-day concentration range from 80 to 130 years.

Predicted rises in average global temperatures for a number of emission scenarios published in the IPCC Report (Houghton *et al.* 1990) are shown in Table 3.3. These are computed from a simplified atmosphere–ocean model calibrated against the GDFL global circulation model described above, to give a 2.5 K rise when in equilibrium with a doubled CO_2 concentration.

SEA-LEVEL RISE

A potentially important consequence of greenhouse warming is the melting of sea-ice and ice sheets on land, but only the latter results in a rise in sea-level. The sea-level will also rise as the ocean waters expand in response to the additional warming. Estimates of these consequences involve large uncertainties because of lack of observations and of understanding of the mass balance and dynamics of glaciers and ice sheets. Moreover, there is considerable uncertainty in the predicted increases in surface temperature due to greenhouse warming.

Glaciers and small ice-caps are very small in volume compared with the major ice sheets, but are liable to melt much more rapidly. Their melting is calculated to have contributed about 40 per cent to the total sea-level rise of some 10 cm over the last 100 years (see Table 3.4).

Table 3.3 Future predictions of CO_2 concentrations, rises in global temperature, and sea level for various emission scenarios

(a) *Business-as-usual, high emission scenario*
   CO_2 emissions continue to increase linearly with time adding 2% of 1990 value each year. Montreal Agreement 75% implemented CH_4 and N_2O continue to increase at 1990 rates

	1990	2030	2060	2100 AD
CO_2 concentration	354	470	590	850
ΔT (K)	0	1.1	2.0	3.25
Sea-level rise (cm)		18	38	65

(b) *All emissions kept constant at 1990 rates*

CO_2 concentration	354	420	465	520
ΔT (K)	0	0.72	1.1	1.6
Sea-level rise	0	15	27	42

(c) *2% p.a. (compound) reduction is all emissions from 1990*

CO_2 concentration	354	388	395	390
ΔT (K)	0	0.4	0.4	0.3
Sea-level rise	0	11	18	21

(d) *2% p.a. increase in all gas emissions 1990-2010, thereafter a 2% p.a. decrease*

CO_2 concentration	354	436	458	464
ΔT (K)	0	0.93	1.10	1.0
Sea-level rise	0	17	28	34

The mass balance of the great Antarctic ice sheet, and of the Greenland ice sheet with only one-tenth the volume, are determined by the difference between accumulated snowfall on the one hand, and melting and calving on the other.* In Antarctica, observations suggest that this accumulation is very nearly balanced by calving of slabs of ice on the ice shelves with little melting taking place because of the very low air temperature. In Greenland, accumulation is balanced about equally by melting and calving. In neither case is there any direct evidence that the ice sheets are far from equilibrium, so that together they are

---

* 'Calving' is the breaking off of large masses of ice from the ice shelves. This ice then forms a part of the ice packs which surround Antarctica and Greenland.

Table 3.4 Estimated contributions to sea-level rise over last 100 years (in cm)

	Low	Best estimate	High
Thermal expansion	2	4	6
Mountain glaciers	1.5	4	7
Greenland ice sheet	1	2.5	4
Antarctic ice sheet	−5	0	5
Total	−0.5	10.5	22
Observed	10	15	20

(*Source:* IPCC Report: p. 274)

unlikely to have contributed more than 20 per cent to sea-level rise over the last century.

The remaining 40 per cent of this rise is attributed to thermal expansion of sea-water. Since this is very sensitive to temperature, being six times greater at 25 °C than at 0 °C, the rise in sea-level depends very much on the depth to which the warming penetrates and therefore the mass of water which expands. This can, in principle, be determined from a fully coupled atmosphere–ocean model, but no such long-period calculations have yet been made. Calculations based on a simple one-dimensional model and quoted in the IPCC Report (Houghton *et al.* 1990), indicate that thermal expansion has contributed 2–6 cm, with a best estimate of 4 cm, to a total estimated rise of 10.5 cm (Table 3.4).

If atmospheric and surface temperatures increase due to greenhouse warming, thermal expansion of the oceans and melting glaciers are likely to continue to make the largest contributions to sea-level rise. Table 3.5 shows that the best estimates of contributions for a temperature rise of 1.1 K by the year 2030 are 10 and 7 cm, respectively, to a total rise of 18 cm. If by 2060 the global warming increases to 2.0 K, the corresponding sea-level rise is estimated to be 38 cm; by 2100 it may rise by 65 cm in response to a temperature increase of 3.25 K. Estimates for lower emission scenarios leading to slower temperature increases are given in Table 3.3.

The greenhouse effect and global warming

Table 3.5 Estimates of contributions to sea-level rise (in cm) from 1990–2030 according to business-as-usual scenario in Table 3.3.

	Low	Best estimate	High
Thermal expansion	6.8	10.1	14.9
Mountain glaciers	2.3	7.0	10.3
Greenland	0.5	1.8	3.7
Antarctic	−0.8	−0.6	0
Total	8.8	18.3	28.9

(*Source:* IPCC Report: p. 276)

All these estimates of sea-level rise, which are probably uncertain by a factor of two, are much less than exaggerated claims based on the assumption that the western Antarctic ice sheet will largely disintegrate and melt. Most glaciologists discount such a scenario, rendered even more unlikely by the recent model climate simulations described above which produce very little greenhouse warming in Antarctica.

CONCLUSION

It is virtually certain that the troposphere is warming very slowly in response to the continually increasing concentrations of CO_2 and other 'greenhouse' gases, but the signal is as yet too small to detect above the large natural climate variations, partly because it is being delayed by the thermal inertia of the oceans. Predictions of the magnitude and timing of the greenhouse warming and of the concomitant changes in rainfall and other climate parameters, come entirely from physico-mathematical models of the global climate system. Unfortunately, the differences between the various model predictions, which are very sensitive to how clouds and their interaction with the radiation fields are represented, are too large to provide firm guidance for major policy decisions. Continued improvement in model resolution and model physics should cause the predictions to converge and thereby narrow the range of uncertainty. This will require several years of model development, especially in respect of the oceans; much faster computers; and, above all, an adequate supply of global observations from both the atmosphere and the oceans to feed and validate the models, and to monitor the actual changes in climate that may eventually become evident.

In the meantime, although the current best estimates of global warming are not so alarming as to warrant *major strategic* changes in energy use, agriculture, and so on, industrialized countries should take all reasonable and practicable steps to restrain or reduce energy consumption, utilize all fuels more efficiently, and explore economically promising alternatives to fossil fuels. The decision of the British Government to restrict UK emissions of CO_2 by 2005 to current levels and to reduce emissions of CFCs in line with the Montreal Agreement are realistic first steps.

In addition, we should develop without delay, adaptive strategies in agriculture, forestry, coastal defences, water supply, and so on, to make the economy less vulnerable to climatic changes when they occur. It would appear that we have a breathing space of some 50 years but this may prove too optimistic; in any case, it is none too long.

ACKNOWLEDGEMENT

This written version of the lecture is largely based on an article entitled 'The Greenhouse Effect' published in *Contemporary Physics* in November 1989, now extended and brought up to date. Figures 3.1–3.7 are reproduced from that article, with permission from the publishers.

REFERENCES

Houghton, J. T., Jenkins, G. J., and Ephraums, J. J. (eds) (1990). *Climate change—the IPCC scientific assessment.* Cambridge University Press.

Manabe, S., Bryan, K., and Spelman, M. J. (1990). *Journal of Physical Oceanography,* **20,** 722.

Mason, B. J. (1987). The greenhouse effect. *Contemporary Physics,* **28,** 49.

Mitchell, J. F. B., Senior, C. A., and Ingram, W. S. (1989). *Nature,* **341,** 132.

Wilson, C. A. and Mitchell, J. F. B. (1987). *Journal of Geophysical Research,* **92,** 1331S.

4
Implications of global climatic change
Crispin Tickell

Sir Crispin Tickell, Warden of Green College, Oxford since October 1990 brought to the lecture theatre an unusual combination of international experience and specialist knowledge of the environment. His distinguished career in the Diplomatic Service culminated in the posts of British Ambassador to Mexico, Permanent Secretary of the Overseas Development Administration (the government department responsible for aid and technical assistance), and British Permanent Representative (Ambassador) to the United Nations, a position which he held until taking up his present appointment in Oxford. His last two official appointments, in particular, enabled him both to expand and deploy the environmental interests which he had developed in parallel with his diplomatic career. During a sabbatical year at Harvard University in 1975–6 he wrote a study of Climatic change and world affairs *which was well received, on publication, by expert and layman alike. His knowledge of climatology and imaginative interest in the social implications of climatic change were acknowledged by his appointment as an unofficial adviser on environmental matters to Margaret Thatcher. He delivered his lecture on the day of her resignation as Prime Minister.*

The prospects of climate change have become of major concern. Battle has been joined between politicians, economists, businessmen, sociologists and others, as well as scientists. There have been debates in the United Nations General Assembly, a multiplicity of seminars, meetings, and conferences culminating in the Second World Climate Conference at Geneva in November 1990, large numbers of books, reports, and articles, and a general spread of unease that the comforting familiarities of weather and climate could be drastically changed to the common disadvantage.

Implications of global climatic change

Sir John Mason gave the third Linacre Lecture on the scientific aspects of climate change and I will not repeat his main points. It is sufficient to say that climate change and its myriad implications are now on the world's agenda.

The subject has long been close to the heart of Margaret Thatcher who resigned as Prime Minister on the day this lecture was delivered.* I first became aware of her interest when, as one of her summit assistants (or sherpas), I helped put together a British initiative on the environment for the London Economic Summit of 1984. I have been lucky to work on it with her in different capacities in more recent times when she placed it with her usual vigour on the highest political agenda.

Her Royal Society speech of 27 September 1988 was in retrospect one of those events which from time to time make people think differently from the way they thought before. There is no way back on such occasions. This speech was followed by an all-day seminar at 10 Downing Street on 26 April 1989. Key members of the Cabinet and leaders of the business as well as scientific communities attended. It fell to me to present the Government's ideas to the world at large in a speech to the United Nations Economic and Social Council in New York 12 days later.

On 8 November 1989 Mrs Thatcher flew to New York to speak to the General Assembly on this subject alone, so far as I know the only head of government ever to have done so. Intensive work followed on the Government's White Paper on the Environment, published in September 1990. At the beginning of November 1990 Mrs Thatcher addressed the Second World Climate Conference in Geneva in another remarkable speech on the implications of climate change for the world community and the actions which individual governments should take.

Few if any in her position can have shown such zest in such a cause. To it she brought intellectual curiosity, an ability to establish a vital but unfamiliar element in politics, economics, and much else, and above all skill in bringing politics to the

---

* 22 November 1990.

service of science in what must surely be the common human interest.

Any discussion of climate change should remark on the smallness of man, and the vastness of the sky above him. Climate is invariable only in its variability. We are used to changes in the weather within broad limits, but we are unused to those more drastic changes of climate which have taken place from time to time during the 4.5 billion years of the earth's history. Indeed we are still in a somewhat abnormal epoch which began around 2.5 million years ago with the onset of the ice ages. In times past people thought there had been four such ages, and that the last one was safely behind us. In fact there seem to have been between 20 and 30 at roughly one hundred thousand year intervals, interspersed with warmer moments of between 10 000 and 15 000 years. We are now well into the second half of our warm period, and there are reasons to believe that the world will cool in the next 4–5000 years. In the warm period which began around 10 000 years ago, there has been all human civilization. Societies based on agriculture began around half way through the period, and the industrial society we now know began only 250 years ago. So far all human societies have crashed, and ours may eventually suffer the same fate.

Industrial society has changed the face of the earth. It is based not only on the technical skills on which we pride ourselves, but also on unprecedented consumption of natural resources, especially fossil fuel in the form of coal, oil, and natural gas. All energy comes from the sun, and fossil fuel is our reserve of stored sunlight, which we have been using at an accelerating rate.

There are four obvious effects of the industrial revolution. The first has been a vertiginous increase in human numbers: from around two billion people in 1930, the year of my birth, to well over five billion in 1990. Short of some major disaster, the population will have risen again to over eight billion in the next 40 years. Secondly, and linked with human population increase, there has been a steady degradation of the land surface of the earth (somewhat less than 30 per cent of the surface as a whole). Of the land surface, only part is suitable for growing crops, and a third of that is already in various stages of desertification. Thirdly, we have polluted most of our rivers, and the coastal waters of the sea. Pollution is spreading to the deeper oceans at a steady rate. Last, we are changing, and again polluting, the chemical character of the atmosphere. Three specific problems are conspicuous. Acid precipitation is regional in scope and soluble if governments have a mind to it. Then there is ozone depletion caused by man-made chlorofluorocarbons. This problem is global in scope but also soluble with political will. Thirdly, the increase in the quantity of greenhouse gases could lead to major changes in the world's weather and climate. Climate knows no boundaries and the problem is one which affects the world as a whole. It is also of mind-defying complexity.

Concern about global warming goes back almost 200 years. In the early 1970s some prophesied an early return to the ice age, and others a hot-house earth. Now we have a broad scientific synthesis on which to work in the form of a Report by the Intergovernmental Panel on Climate Change, in particular the conclusions of a working group under the chairmanship of Dr John Houghton, Sir John Mason's successor at the Meteorological Office. The Report of the Panel was approved at the Second World Climate Conference, and can be regarded as the best that scientists can do in an evolving science. In parallel, a Panel of the US National Academy of Sciences has been considering the policy implications of the likely global warming produced by the greenhouse gases, and its Report will be published in the United States in 1991.

Let me summarize briefly the main conclusions. Greenhouse gases already make the earth 33°C warmer than it would otherwise be. On the assumption of business-as-usual (in other words we carry on as we are), the average global mean temperature will rise by around 0.3°C a decade, in short an increase of 1.0°C by 2025, and 3.0°C before the end of the next century. In case anyone should think that a rise of around 1.5°C above that prevailing in pre-industrial times is small, we should bear in mind that the average temperature was only about 4°C less during the last ice age, and that in a world mostly covered by water, the land will warm up much more than the sea.

Again on the assumption of business-as-usual, there would be wide regional variations. There would be more precipitation generally, and considerable changes in patterns of rainfall. But if certain areas received more water, others would receive less, particularly in summer; and Southern Europe and North America could be particularly affected. Snow cover and ice would of course be reduced at the Poles. There is evidence that this is happening already. Last, there would be a rise in sea levels of around 6 cm a decade, or 20 cm by 2030, and up to 65 cm by the end of the century.

There are other important points. One is the long time lag due to the moderating influence of the oceans. Hence the present slow warming of the atmosphere could accelerate as the upper layers of the sea respond. There is thus an important difference between the current temperature which we can observe and the future temperature at which the earth will, as we hope, reach a new equilibrium. If we wish to stabilize greenhouse gases at their present levels, and thereby to halt the inexorable process of warming, we would have to reduce man-made emissions of greenhouse gases by around 60 per cent. No one is even thinking of reductions of this magnitude. Indeed the United States, as the main producer of greenhouse gases, so far refuses to consider any reductions at all.

There are many uncertainties. We do not fully understand the role of clouds and the hydrological cycle, nor the role of the oceans. We have a theoretical understanding of the carbon cycle, but cannot yet say how all the carbon we are adding to the atmosphere is absorbed. Something like 40 per cent is unaccounted for. Nor do we understand the behaviour of the polar ice sheets, and the annual spread and contraction of sea-ice. Least of all do we understand the likely reaction of ecosystems, including our own species, to rapid climate change. There are of course other factors. Volcanic emissions have immediate effects on climate. There could also be disasters of a kind not easily foreseen. Three examples are: major changes in the ice now ashore on the Antarctic continent; changes in the system of ocean currents which keep England a green and pleasant land; and desertification in the interiors of continents as has been seen often before in the earth's history.

If ever there was a global problem, this is it. Indeed, it usually has a paralysing effect on the minds of those who make policy, partly because the changes are on a time-scale that is rapid by geological standards but slow by those to which we are accustomed, and partly because the effects could bring into question the fundamentals of our society. It is too late to stop the process. The best we can do is to see how and where the effects might be mitigated and how and where we can best adapt to them. These will be the focus of the forthcoming Report of the US National Academy of Sciences.

Any action we might choose to take should be based on four main considerations. Granted the continuing uncertainties in the science, and the disruption which drastic action would do to our economies, governments should consider doing now what makes sense for reasons other than climate change on its own. Next, they should consider pre-emptive action of a modest kind, rather like paying the premium on an insurance policy against relatively unlikely disaster. Then they should redirect and give more financial support to research into the main areas of uncertainty. Here a co-ordinated international effort is essential if duplication and waste are not to take place. One of the recommendations of the Second World Climate Conference was the creation of a global climate observing system. It is extraordinary that at present observational work is deteriorating in quality, especially in the southern hemisphere. Finally, we should never forget that climate change is only one of the hazards facing us in the future, and not perhaps the worst. Any action we take must always be seen in a wider framework of problems involving over-population and environmental degradation generally.

Such action falls into three main areas. Obviously it is our use of energy in the form of fossil fuel which has most increased greenhouse gases in the atmosphere. Cutting down carbon emissions to the necessary levels would require major changes in the way we generate and use energy. Whether climate changes or not, we need to conserve energy, increase energy efficiency, find more economical methods of transmitting power, and look for building designs and domestic practices which minimize waste. We need to develop alternative sources of energy, in particular solar energy, so as to reduce dependence on fossil fuels. Such sources are well known. Many carry hazards of their own. If governments had spent as much on solar energy in the last 50 years as they did on developing nuclear energy, either fission or fusion, the world would now be a different place. We need to look at new transport systems, from systems of mass transport to vehicles for individuals. In an age of information technology we need to look at the ways in which we organize our work, decide whether daily mass movements of people in and out of cities is desirable, and consider what kind of urban infrastructure best suits our requirements. In so doing, we have to reckon with the problems of biodiversity, or the conservation of other species and habitats in the system of life of which we are no more than a tiny part.

Next we have to look at ways in which we use the land. Forest destruction represents a holocaust of the genetic riches of the earth. This destruction is another source of the greenhouse gases. Deforestation can also lead to climate change. We need to look at the balance between trees and crops, inhabited and uninhabited land, and human and industrial fresh-water requirements. In my judgement, it would be as well if all Ministries of Agriculture and Forestry Commissions were abolished and instead we had Ministries of Land Use, which included town and country planning where problems of land use and water management could be seen together. In the light of possible future floods or droughts, we also need to look at new agricultural methods. Some plants would benefit from a warmer world with more atmospheric carbon dioxide, while others would suffer. If the sea levels rose, we would have to look to our coastal defences, bridges, oil platforms, inshore aquefers, and fisheries.

Finally, the processes of industrial society need new scrutiny, not only those causing chemical emissions and pollution, but also the siting and shape of industry and the disposal of its many wastes. The growing use of information technology will be important in this respect.

Taken together, this is a bill of action which far exceeds the capacities of most governments. The first instinct of many is to hope and look for evidence that the changes forecast will not happen, or at least are exaggerated. Scientists can sometimes be brought in from other disciplines to express scepticism dressed up as common sense. Not surprisingly, many have looked at the business-as-usual predictions as those most likely to happen. Relatively few have looked at the worst-case predictions. These need more attention than they have so far received.

Some governments have sought to shelter behind the obvious need for international action as an excuse for not taking decisions at home. International action is by its nature difficult and contentious. But during the last five years, and with the IPCC Report and the Second World Climate Conference, there has been increasing pressure for results. In some countries the leadership has come from governments and in others from public opinion. The UN Conference on Environment and Development (UNCED) in June 1992 will be the next major event along this path.

Already the countdown to the conference has begun. There are hopes for a climate convention to lay down a code of good climatic behaviour. This would follow the precedent of the Vienna convention and protocols on ozone depletion rather than the conference on the Law of the Sea which tried to do too much too quickly. A convention on climate could provide a framework in which detailed protocols on specific greenhouse gases could later be fitted. There are also hopes for a convention on biodiversity, or the protection of threatened ecosystems and species, a convention or protocol on forestry, and an agenda for action (Agenda 21) reaching into the next century. In this way global problems can be dealt with, each one on its merits, but all taking account of each other.

There has been discussion, particularly among diplomats, about what institutions and instruments will be necessary to deal with these unprecedented problems. Obviously certain technical institutions already exist: the United Nations Environment Programme, the World Meteorological Organization, the World Climate Programme, and the Intergovernmental Panel on Climate Change. But work on the environment, and by implication climate, is also covered by others among the 40 or so UN agencies and associated bodies, and a prime need in the future will be to establish better co-ordination between them.

Something more central and political may also be required. The UN General Assembly, the UN Security Council, and the International Court of Justice could all serve useful functions in co-ordinating international action. But there are arguments for something more specific in the form of a new international authority, an Ecological Security Council, or even an adaptation of the idea of the Baruch Commission (set up in 1945 to look into the then new problems of nuclear energy). My own view is that if we could agree on conventions in 1992 governing the climate, biodiversity, forestry, and other environmental issues, we should, almost without realizing it, be creating a base for something like the General Agreement on Tariffs and Trade with its disputes procedures, reconciliation mechanisms, countervailing measures and the rest, for use if required. Thus we could have a system which could grow and respond to need, rather than a brand new institution which, assuming it could be set up, could all too easily prove inflexible and become out of date.

In any international system we must recognize the needs of equity, a point well brought out at the Second World Climate Conference. Unwittingly or not, the industrial countries have made most of the mess, with 75 per cent of current emissions of carbon dioxide. Although the poor and more vulnerable countries would probably be worst affected in a warmer world, the industrial countries would be affected too, some of them severely. They need to give an example and take the lead.

Without such an example, poorer countries might not be inclined to take their warnings seriously. At present the average American puts more than four times the world average of carbon or its equivalent into the atmosphere every year, and more than ten times the contribution of the average Chinese. Yet the United States constantly expresses alarm at the prospect that China will rely on its low-grade coal for its future economic development.

In practical terms the best guide to willingness to join in international action is identification of interest. Every government will have to recognize a national as well as international interest. Problems range across the whole international spectrum: from the ever-hesitant United States, the chaotic Soviet Union and eastern Europe, and the not-so-efficient but well-meaning members of the European Community on the one hand, to the big oil producers, the big forest burners, India, China, and those nations so poor that they cannot consider any changes without external help on the other.

So far it has been some governments and many individuals in the industrial countries who have raised most of the alarm. Against them have been ranged huge powers of inertia, reluctance to face the implications of change, and in poor countries the suspicion that protection of the environment is only a luxury of the rich who would like to use environmental standards to hold them back. Yet in logic it should be the poorer countries as the most likely victims of change who should be most vigorously calling for international action. This may well happen as scientists in such countries, in particular China and India, correctly measure the costs of inaction. It may also happen as a much neglected international problem, that of refugees, becomes more prominent.

In the past people could move when the environment became adverse. Now the doors are closed. In 1978 there were roughly five million refugees on a strictly political definition. In 1990 there were almost 15 million. If we widen the definition to include environmental refugees, the total could come out at around 25 million. Yet in the future that figure could be multiplied several times. With changes in patterns of rainfall and sea-level rise, combined with environment degradation and population increase, especially in poor countries, we could find that displaced people could swamp existing boundaries and make substantially worse the other political and economic problems caused by change.

Whoever takes the lead in pressing for international arrangements, national governments still have major responsibilities. Some of these were brought out in the British Government's White Paper entitled *This common inheritance* published in September 1990. The language of the White Paper was less strong than it might have been, but at least it fixed the vital principle that policy required co-ordination at the centre and that the environmental dimension should be established in all activities of government.

The White Paper was only a beginning. Yet most governments in industrial countries have not yet come so far. All governments will eventually find themselves obliged to undertake environmental costing, to bring the environment into the national accounts, to fix fiscal incentives and disincentives, and generally to set new ground-rules for the operation of the market. Little could be more important than rejigging the price mechanism. Two examples will suffice. For how long can governments afford to favour private transport, in particular cars, at the expense of the public transport systems? For how long can governments connive at a system of energy pricing which, for instance, permits coal to be sold for a price which fails to take account of the damage caused by its combustion? The application of the 'polluter pays' principle could have implications far beyond what most governments suppose.

In seeking to understand and cope with this huge complex of problems, we have to remember three main points. First, we are not coping with one factor but with a combination of factors, of which climate change is only one. Secondly, science is full of surprises. Consensus is temporary and relative. Physical mechanisms do not proceed in linear fashion. There may be unknown mechanisms to correct global change, or others to make it worse. Last, the problem is not change itself but the rate of change in an overcrowded world.

It may not be a consolation, but even if we warm up the earth to degrees unprecedented in recent times, and even if we damage the circumstances which made our species dominant, we should remember that the ice may well return in 3000 to 4000 years' time with prospects of full glaciation in 20 000 years time. Life itself is in no danger. It can carry on without us as it has done for all but a minute fraction of the history of the earth.

REFERENCES

Department for the Environment (1990). *This common inheritance: Britain's environmental strategy.* White Paper 0101120028.

Houghton, J. T., Jenkins, G. J., and Ephraums, J. J. (eds) (1990). *Climate Change: The IPCC scientific assessment.* Intergovernmental Panel on Climate Change. Cambridge University Press.

5
The earth is not fragile
James E. Lovelock

Professor James Lovelock, FRS, is one of the most eminent representatives of that rare breed, the independent scientist. The first 20 years of his career were spent with the National Institute for Medical Research in London; during the same period he held visiting Fellowships at both Harvard and Yale Universities. In 1961 he was appointed Professor of Chemistry at Baylor University's College of Medicine and began an important series of consultancies to NASA which led to his pioneering work in measuring chlorofluorocarbon (CFC) gases in the atmosphere and predicting their effects. In 1964 James Lovelock broke free from the constraints of institutional science and pursued his research independently: 'You may well ask', he wrote in the introduction to his book The ages of Gaia, *'whatever became of those colourful romantic figures, the mad professors, the Drs Who? Scientists who seemed to be free to range over all of the disciplines of science without let or hindrance? They still exist and in some ways I am writing as a member of their rare and endangered species'. James Lovelock used his freedom to develop the 'Gaia theory' which he expounded in this lecture and which was first formulated in his book* Gaia, *published in 1979. Ten years later,* The Independent *newspaper, in a leading article, commented that in his address to the 'Friends of the Earth' in September 1989 'Professor James Lovelock slaughtered just about every sacred cow in the scientific establishment and in the environmental movement'.*

Hans Johst once said 'Whenever I hear the word 'culture', I release the safety catch of my Browning!'. I don't know what in particular incensed Johst about culture, but I suspect that he knew the word prefaced a gush of hypocrisy. For me, the same kind of anger comes when I hear that cliché, 'the fragile earth'. I get ready for a flood of words and a televisual presentation of cuddly animals and lush vegetation. The words may come from a prominent 'green' presenter, but to me they will sound green only with the mildew of insincerity. In Victorian times that same word 'fragile' was used with a similar tendency to describe women, and to justify their domination. They were called fragile because it implied a feeble delicacy that needed male protection—Victorian women were not fragile, they were tough, they had to be to survive.

So it is with the earth. Fortunately for us it is very tough indeed. In the nearly four billion years of its existence as a live planet, it has survived at least 30 major planetesimal impacts, each of them devastating enough to destroy more than half of the life present. In addition, solar output has increased by 25 per cent and perturbations have occurred, such as the appearance of oxygen as the dominant chemical species. What we are doing now in the way of pollution and destruction of natural ecosystems is by comparison a minor upset. Those who call the earth fragile, or who say that some human act will destroy all life on earth, are either ignorant of what the earth really is or are using 'earth' metaphorically as a synonym for humans. Either way we use 'fragile' as did those Victorian men about their women: applying to our planet a dependent status almost as if it were a possession.

We still talk about the earth as if the planet in our minds were a multicoloured political sphere mapping the territories of tribes and nations. The real earth, that stunning blue and white sphere, has become just a visual cliché, no longer inspirational, a banal image advertising soap on satellite television.

In this lecture, I want to discuss our relationship with the earth. In particular, I would like to put before you some thoughts on the consequences of taking on the responsibility for the management, or if you prefer, the stewardship, of the earth.

As the only organized intelligence, perhaps we have the duty as well as the right to take charge of the earth and govern it responsibly. Maybe so, but first we must ask, what is the earth? This may seem a trivial question—everyone knows what the earth is—but unfortunately there seems to be no common view. It is almost as if the earth were an evolutionary inheritance, which we, like fleas on the back of a camel, just take for granted, never noticing what it is. Even scientists differ about what the earth is.

There are three scientific views of our planet. First, a small minority that includes me, who call ourselves geophysiologists, see the earth as a quasi-living system, or if you prefer, a planetary sized ecosystem—something called Gaia. We postulate that this system automatically regulates such important properties as climate and atmospheric composition, so that they are always more or less comfortable for life.

A larger minority prefer what the climatologist Stephen Schneider has called co-evolution (Schneider and Londer 1984). They see life and its environment as only loosely coupled. They agree with geophysiologists that the composition of the air, the ocean, and the rocks is affected by the presence of life, but they reject the idea that the earth may be self-regulating in such a way as to sustain a comfortable environment.

Co-evolution originated from the ideas of the Russian scientist Vernadsky. He was the man who first used the notion of the biosphere, in the familiar vague way we still use it (Vernadsky 1986). Co-evolutionists recognize the need for interdisciplinary research and are the force behind the Global Change and the International Geosphere Biosphere Programs. It may be some time before we know whether co-evolution or Gaia is nearer the truth.

There are a few, mostly geographers, who see the earth as a whole, but the majority of scientists, even if they give lip-service to either Gaia or co-evolution, still act as if the earth were a ball of white-hot, partially melted rock with just a cool crust moistened by the oceans. On the surface they see a thin green scum of life whose organisms have simply adapted to the material conditions of the planet. With such a view go metaphors like 'the spaceship earth': as if humans were the crew and the passengers of a rocky ship forever travelling an inner circle around the Sun; as if the four billion years life has existed on earth were just to serve as our life-support system when we happened to come aboard. Seen this way, obviously the earth might appear fragile, like one of those great greenhouses, called 'biospheres', in Arizona. Those who so see it must wonder how it has survived so long.

This is the conventional wisdom about the earth, and is still taught in most schools and universities. It is almost certainly wrong and has arisen as an accidental consequence of the fragmentation of science, a fragmentation into a growing collection of independent scientific specialties. Practising scientists are aware of the limitations of this diffuse conventional wisdom about the earth, but even when they are specialists in some branch of earth or life science, they still seem to act as if it were true. If we as scientists want to know about life, the universe, or the earth, we read about it in the *New Scientist* or the *Scientific American*. Back in the laboratory, where serious science is done, we continue in our own speciality without concern for either general wisdom or the intricate details of the specialties of our close colleagues.

If you think I exaggerate, try attending discussion meetings on the three closely related earth sciences. For example, on one day you could attend a discussion on the chemistry of stratospheric ozone; on another day, a discussion of the geophysics of fluid motion in the oceans; and on the third day a discussion on the geochemistry of rock weathering. These are all earth science topics but you would find little that was shared in common between them. More seriously, a considerable proportion of the scientists from each of the three discussions would be unaware of the discoveries of the others.

Of course, no single scientific approach can lead to a complete understanding of the earth—all are needed. We need the reductionist model of the earth to understand details at the molecular level. A key example is the chemistry of the stratosphere. It was only through the application of classical atmospheric chemistry and physics that Rowlands and Molina first made known the threat to ozone from the CFCs. From biogeochemistry there came, through the work of G. E. Hutchinson, the recognition of the role of micro-organisms in the soil and the oceans as the source of methane and nitrous oxide. From geophysiology came the recognition that atmospheric gases, like carbon dioxide, methane, and dimethyl sulphide, may be part of a physiological climate regulation.

We are at a time when scientists as professionals seem to have lost sight of the earth as a planet in the intricacies of detail. As a result, when confronted with environmental concerns, they tend to think about specific dangers to people, especially themselves, and ignore hazards that loom on a planetary scale. The foremost personal and public fear is that of cancer. Consequently, any environmental chemical or radiation thought to cause cancer is given attention out of all proportion to the real risk it poses. Nuclear power, ozone depletion, and chemicals like dioxin and PCBs, are regarded as the most serious of environmental hazards because of this fear, but also because nuclear radiation and halocarbons are so easy to measure. I think that the potential hazards of the gaseous greenhouse and land abuse have, until recently, been ignored because they perturb the planet, not individual people, and because they are much more difficult to quantify.

Sir John Mason in his lecture discussed the magnificent numerical models of the atmosphere that he and his colleagues have used to advance the science of weather forecasting. He also described how similar models could predict the climatic effects of greenhouse gas accumulation. He would be the first to admit that these models of the atmosphere are far from able to include the full effects of clouds. He would confirm that they are still less able to include the effects of the internal climate of the oceans, and almost entirely unable to take in the changing and responding ecosystems at the surface of the earth.

So how are we to know what the earth is as a system? How can we govern it if science is still decades away from telling us what it is? Should we wait for the deliberations of the plenary session of the 'all-science interdisciplinary congress'? Or should we listen to thoughtful environmentalists, like Jonathan Porritt, who ask if we can afford to wait for scientific certainty before taking action on environmental affairs.

Consider for example the accumulation of chlorofluorocarbons (CFCs) in the atmosphere. No one doubts that the CFCs have reached a level that is already damaging. We are all agreed that the emissions of the long-lived CFCs should be banned immediately.

Sensible 'Greens' are puzzled about why if this is so, we continue to spend billions of scarce funds on stratospheric and ozone depletion research, when the problem is in effect solved. We know the poison, all that needs doing is to stop imbibing it. If we were serious, say the Greens, we should be considering, in addition to a world-wide ban on the manufacture of long-lived CFCs, the general problem of their uses. How can we refrigerate and air condition without letting loose CFCs to the atmosphere? How do we dispose of the large stocks of CFCs in storage and in the refrigerators now in use? Compared with the excitement and glamour of research in the upper atmosphere, or of exquisite physical chemistry experiments, or of elegant computer models, CFC disposal is a problem for mere engineers, the rude mechanicals of science. Proof of the lack of this kind of simple engineering is the fact that a substantial proportion, possibly more than half of the CFCs entering the atmosphere, come from leaking American car air-conditioners. The Greens are right. Engineering research urgently needs doing and should be at the top of our lists for action instead of at the bottom.

We don't need to burn a billion candles to specialist scientific research, it only pollutes the air and in the end may do nothing but confirm that it is too late to postpone our doom. Let us spend our cash now, with the planet rather than people in mind. Otherwise we are like a farmer who would mortgage his land to pay for his children's education without realizing that in so doing he denied them their inheritance.

In the last century the Victorians were faced with environmental problems just as serious as our own today. In the mid-nineteenth century there were epidemics of the water-borne diseases cholera and typhoid that caused the death of a third of the inhabitants of a city in a few months. Many prayed for deliverance but there is no report of any success with this venture. Science was not then organized as a powerful lobby and was prepared to admit that it did not know the cause of the diseases. Physicians at the sharp end of this battle suspected, from the epidemiology of the diseases, that infection was waterborne or came from the bad odours of the primitive sewerage systems then in use.

Our sensible forefathers did not pour funds into the infant science of microbiology and wait until it proved that cholera and typhoid were water-borne bacterial infections. They acted promptly and empirically by installing clean water supplies and efficient sewage collection and disposal plants. Engineering was in those days a proud profession and triumphantly displayed its self-confidence in those amazing Gothic pumping stations, now a place of pilgrimage for students of architecture.

But the Green movement itself is a potent force preventing environmental reform. It has anachronistic views of industry, which it regards as harmful and polluting. The movement includes many who condemn and dislike both science and technology. Some Green philosophers and leaders are refugees from the older humanist movements. Among them are those primitives who see all industry as inherently bad and existing solely to benefit profiteers. Others are William Morris groupies who would return to a romantic but impractical rural existence. This kind of 'green' nonsense is encouraged by the tendency of talented writers and dramatists to cast their villains as owners, or employees, of the nuclear or chemical industries. They go on to make these industries the stage equivalent of the desecrated graveyard of a Victorian melodrama. The nuclear industry is all too often seen as a place of quintessential evil. This flood of 'green' propaganda ignores the certainty that if we gave up our industrial civilization, only a few of us would survive. More seriously, these story-tellers deny the possibility that industry could reform and become non-polluting and benign, and as always their concern is for people not for the planet.

We could go along with the more responsible among the Greens, and still retain the long-term guidance of science, if we could first delineate the near certainties and then act empirically in an engineering way. We now know, for instance, from the record of the gases trapped in the layers of Antarctic ice, just how the atmosphere changed during the past 200 000 years. We know for certain that the carbon dioxide 15 000 years ago in the depths of the ice age was about 180 parts per million and that at all times there has been a close correlation between carbon dioxide and temperature. The rise in temperature and the rise in carbon dioxide over a period of only a few hundred years close to 12 000 BP, was 3 °C and 100 parts per million respectively. In the past 200 years we have increased the atmospheric carbon dioxide by 80 ppm, and if we take the greenhouse effect of the CFCs into account, we have already effectively increased the greenhouse effect by as much as happened between the ice age and the interglacial that followed. We are near certain that the mean global temperature will rise, even if we stopped emissions now. The uncertainty is less about the rise in temperature than about how fast it will take place.

If we emulate our Victorian ancestors and develop an empirical approach to planetary problems, we might find it helpful, even if only notionally, to introduce a new profession, that of planetary medicine. It would stand to specialist earth science in the same way that medicine stands to biochemistry and microbiology and would be the environmental equivalent of the practice of medicine. Its general practitioners would be concerned with the health of the planet and an important part of their practice would be planetary preventative medicine.

For example, let's consider the production of electricity. Fossil-fuel-burning power stations are among the principal atmospheric sources of carbon dioxide, sulphur dioxide, and nitrogen oxides. They are at best only 40 per cent efficient and waste the rest of the energy of the fuel in those ugly cooling towers that advertise their presence. Attempts to remove sulphur oxides from the effluent of these power stations, so as to reduce acid rain in distant places, would be seen by a planetary physician as treating a symptom only. The right therapy would be to burn coal in a way that made available most of the energy of combustion and so that all the pollutant emissions were gathered and either used profitably or stored where they would do no harm. I have been told by Shell that pilot plants in which coal is converted to hydrogen and carbon dioxide already exist. In these plants the hydrogen would be the fuel of gas turbine power stations and the carbon dioxide would be collected and disposed of underground or in the ocean. The UK coal industry have proposed a similar scheme. At present, although more efficient in energy terms, these alternative power sources are not yet competitive in cost with conventional power stations. But no sensible engineer would expect pilot plants to compare in efficiency with the evolutionary product of a long-running commercial operation. Also the cost of pollution is not yet charged against the power producers. A change in attitude could make these non-polluting ways of burning fossil fuel in fact less expensive than conventional power stations.

It will not be easy to convert industry to supplying and distributing alternate fuels such as methanol or hydrogen for transport, but by so doing the emissions from road vehicles and aircraft could be greatly reduced, even eliminated.

Another way for the broader vision of a planetary physician to help would be through re-examining that 20-year-old photograph of the earth from space—the one that showed our planet as beautiful and seemly when seen in its entirety. This picture of the earth would remind us that 70 per cent of the surface is ocean, something often forgotten in our obsession with human affairs on the land surface.

We are just beginning to glimpse the extent to which the world's oceans and the marine life in them are important in regulating the climate and the chemistry of the earth. Marine biology is a Cinderella among sciences, particularly in the UK. Yet it was the pioneering work of my colleagues Patrick Holligan, Andrew Watson, and Mike Whitfield at the Plymouth laboratories, that first drew attention to the planetary significance of microscopic algae living at the ocean surface (Holligan *et al.* 1983).

The algal life of the oceans is important for climate in three possible ways. Firstly, it emits sulphur gases that oxidize to form cloud condensation nuclei. The clouds over the oceans may therefore be a consequence of the organisms that live in the sea surface. It may be some time before the importance of this observation for global climate is established. But clouds do seem to have a net cooling effect, and Andrew Watson drew my attention to the fact that the long-term average of cloud distribution seen from space shows that the clouds congregate above those areas of ocean rich in algal growth.

The second discovery concerning marine algae came from the observations made by British scientists in the North Atlantic last year (Watson *et al.* 1991). They found that the great springtime blooms of the diatoms that cover areas of millions of square kilometres use up the carbon dioxide in the sea surface so that its concentration falls to a level far below that in equilibrium with the atmoshere. They serve as a powerful pump that removes carbon dioxide from the air. Prior to this discovery, geophysicists were confident that the transfer of carbon dioxide from the air to the ocean was controlled by physical forces alone. This is not so surprising since measurements of gas transfer between the air and ocean are most often made by graduate students during summer ocean cruises when algal blooms are rare.

The third way in which algae can affect climate is simply by their presence or absence. In the absence of algae, or if they grow in the form of large organisms like seaweed, the ocean water is so clear that sunlight penetrates deep and does not heat the surface layers. Conversely, the dense growth of microscopic algae absorbs and scatters sunlight and causes surface warming. Satellite photographs of the ocean show large clear areas north and south of the Equator and an increasing algal density towards the northern and southern oceans.

A significant part of these ocean studies involving marine algae has originated and been executed by scientists at Plymouth, and in this research they lead the world. The need for a more open and enlightened approach in environmental science is illustrated by the meagre support received by these scientists. True, there is a national scarcity of money for science, but Plymouth is poor on any comparison. Perhaps it is a reflection of our compassionatre nature as a nation that in science the excellent few are starved to feed the mediocre multitude.

The re-examination of the earth by a planetary physician would reveal the forests of the humid tropics to be another region of the earth's surface with a significant climatic role, and one under threat from people. In spite of optimistic signs coming from Brazil and Colombia, we are still removing tropical forests at a ruthless pace. Yet in the First World, scientists try to justify the preservation of these forests on the feeble grounds that they are the home of rare species of plants and animals, of plants containing drugs that could cure cancer. They may do. But they offer so much more than this. Through their capacity to evaporate vast volumes of water vapour, the forests serve to keep their region cool and moist by wearing a sunshade of white, reflecting clouds and by bringing the rain that sustains them. Their replacement by crude cattle farming could precipitate a disaster for the billions of poor in the Third World. Sir Crispin Tickell in his lecture movingly portrayed the human suffering, the guilt, and the political consequences of a refugee problem involving a major part of the humid tropics. The change in the land surface when the forests went would also have secondary climatic consequences for the temperate regions.

That this danger is real was illustrated in an unusual television documentary about the Panama Canal which was shown a few years ago. The history of this amazing feat of engineering was used to illustrate a new threat to its continued function. The threat was not, as you might imagine, from local politics, but from agriculture. The canal climbs over the isthmus of Panama through a series of locks. The entire system is powered and kept filled with water by the abundant rainfall of that humid region. But the rain and the trees of the forests are part of a single system. Now that the forests are being destroyed to make cattle ranches, the rain is declining and may become too little to sustain and power the canal. I hope that somehow the fact that this great work of man, of engineering, is threatened by our insatiable desire for beef, brought home to viewers the consequences of deforestation.

A planetary physician would see the great fests of the tropics as part of the skin of the earth and, like human skin, they sweat to keep us cool. The tropics are warm, humid, and rainy, an ideal environment for trees, but few seem aware that the trees themselves keep it this way. The wet and cloudy tropics are not a given state of the earth, they form an environment maintained by the ceaseless evaporation of vast volumes of water through the leaves of the trees; the rising water vapour condenses to form the rain clouds that persist above the forest. If the trees are felled the rain ceases and the region turns to scrub or desert. Trees and rain go together as a single system—without the one there cannot be the other. Sweating is part of our personal refrigeration system. The evaporation of water from the forests is part of the cooling system of the earth. It works because the white clouds that persist over the humid tropics have a net cooling effect.

Maybe you think that the forests are so vast that it will take decades to clear them significantly. If you do, you could be wrong. Until recently, an area of forest equal to that of Britain was razed annually. At this rate, in ten years time 65 per cent of all of the forests of the tropics will have gone. When more than 70 per cent of an ecosystem goes, the remainder may be unable to sustain the environment necessary for its survival. To denude the earth of forest is like burning the skin of a human; burns of more than 70 per cent of skin area cannot be survived.

Brazilian scientists were once asked by their government to calculate the value of the forests of the Amazon as producers of oxygen for the world. The government spokesman argued that without the oxygen their trees produced, fuels, like coal and oil, would be worthless. Some charge should therefore be made for the export of the essential gas, oxygen. It was a fine idea, but unfortunately, calculations of the net production of oxygen by the forests gave an answer close to zero. The animals and the micro-organisms of the jungle used up almost all of the oxygen the trees produced.

Amazonia may not be worth much as a source of oxygen, or by the same calculation, as a sink for carbon dioxide, but it is a magnificent air-conditioner, not only for itself but also for the world through its ability to offset, to some extent, the consequences of greenhouse-gas warming. Do the forests have an estimable value as natural, regional if not global, air-conditioners?

One way to value the forest as an air-conditioner would be to assess the annual energy cost of achieving the same amount of cooling mechanically. If the clouds made by the forests reduce the heat flux of sunlight received within their canopies by only 1 per cent, then their cooling effect would require a refrigerator with a cooling power of 14 kilowatts per acre. The energy needed, assuming complete efficiency and no capital outlay, would cost annually £2000 per acre. How does the value of this freely given benefit compare with that of land cleared for cattle raising—which is the usual fate of land in the humid tropics?

An acre of cleared tropical forest is said to yield meat enough for about 750 beefburgers annually, meat worth at the site not more than about £10 and this only during the very few years that the land can support livestock. Unlike cleared land in the temperate regions, beef production cannot be sustained in the tropics and the land soon degenerates to scrub or even desert. Next time you eat a burger, or better, watch someone else eating it, think of the real cost of its production, the stripping of an asset worth £40. Yes, the 55 square feet of land needed to produce enough meat for one burger has lost the world a refrigeration service worth about £40. On this basis an accurate but imprecise estimate of the worth of the refrigeration system that is the whole of Amazonia is about £100 trillion.

Such a valuation in terms of the refrigeration capacity of the trees alone is an underestimate. Just now the forests sustain a home, a habitat for vast numbers of organisms, including a billion people around the earth. The forests are more valuable to us all than we have yet grasped, they are like love itself, something so valuable that we take them for granted.

Common sense now tells us that, in the absence of a clear understanding of the consequences of what we are doing to the earth, we should cut back our pollutions and land abuse to the point where at least there is no annual increase. But like all acts of self-denial it is only too easy to put it off until something happens. I can't arrange, or predict, anything exciting enough to cause us to give up polluting, but what I can do is to tell you a fable about an environmental problem that afflicted an imaginary industrial civilization 15 000 years ago.

Just about the time that our immediate ancestors appeared on the earth, 2.5 million years ago, the planet itself was changing from a state where the climate was comparatively constant to one where the climate cycled periodically between glacial and interglacial phases. The ice ages were long; they lasted some 90 000 years. In the intensely cold winters, ice extended to within 45 degrees of the Equator. The warm periods, the interglacials, were brief, lasting only about 12 000 years and the climate was like the one we now enjoy, or at least enjoyed until quite recently.

I would like you to imagine that civilization became industrial 15 000 years earlier than it did. This requires an increase of only 0.5 per cent in the rate of evolution of human society. I would ask you to envisage a civilization very like ours now but existing 15 000 years ago, just as developed and just as polluted and with lectures like this one in progress. The main difference would be that the earth was then in the cold state of an ice age. The climate here in the position of Oxford would be like the present climate of Iceland and there would be few inhabitants. The oceans would be more than 400 feet lower than now. A vast area of land, mostly near the Equator, that is now under the ocean, would be dry and populated.

Let us imagine that the civilization developed and became industrial somewhere in the region of Japan and China. The cold winters, with the need for housing and heating, stimulated invention. The region was also rich in coal, oil, and mineral deposits. Soon these were exploited and there followed a rapid progression through water and wind power to steam electricity and nuclear power, just as we have seen take place in a mere 200 years.

Greenhouse gases, carbon dioxide, and methane from extensive agriculture began to rise, and before long the climate became perceptibly warmer. A large part of the civilized world was in the tropical zones and these were becoming uncomfortably warm for their inhabitants. The nations of the north were efficient at producing consumer goods. They were like the Victorian British, or the present-day Japanese. Needs drive invention, and soon refrigerators using CFCs were pouring from the production lines and were shipped to customers world-wide.

It was not long before scientists began to realize that the global environment was changing. A few of them stumbled on the fact that CFC gases leaking from refrigerators were accumulating in the air without any apparent means for their removal. Soon it was discovered that the CFCs were a threat to the ozone layer and that if their growth in abundance in the air continued, ozone in the stratosphere would be so depleted that many, especially the fair-skinned, would be in grave danger from the ultra-violet component of sunlight and would develop skin cancers. There was an explosion of hype in the media over this threat and funds flowed for science as never before. Governments were reluctant to act because they knew that CFCs were harmless in the home and were the most efficient refrigerant gases that could be used. And they were the basis of a large and profitable industry. They were reluctant also because there was no evidence of any increase in solar ultra-violet at ground level; indeed, there was a decrease. So nothing was done to stop CFCs rising in abundance at 10 per cent each year.

A few scientists felt frustrated because they knew that the real threat from the CFCs was not ozone depletion but their property of blocking the escape of outgoing heat radiation from the planet. CFCs are more than 10 000 times as potent as greenhouse gases than carbon dioxide. The fear of cancer always seems to transcend other dangers and as a consequence ozone depletion was the issue that received the most attention. Greenhouse warmth was known about but regarded as a good thing, since the world was cold anyway.

Nobody then knew that in 2000 years the planet was destined to make one of its characteristic jumps in temperature to an interglacial. The polar regions in the ice age were so inhospitable that there was none of the clear, strong data from ice cores that we now have to help us understand the past. The jumps in temperature due in 2000 years time would come because the position of the earth with respect to the Sun was changing in a way that increased the heat received from the Sun. These were small increases in solar heating and by themselves insufficient to precipitate an interglacial, but at the end of a glacial period the planet was in a state highly sensitive to small perturbations.

A minority opinion among the scientists held that the increasing warmth from greenhouse-gas pollution would start to melt the ice caps and that the flooding of low-lying tropical forest land would then take place. This in turn would release vast volumes of methane gas as the vegetation rotted beneath a few feet of sea water. The methane would cause more greenhouse warming and soon by a runaway positive feedback the planet would heat and melt the vast polar ice caps. They warned that the rise in sea level would ultimately be 450 feet, enough to drown most of the large towns and cities of the civilized world. Then as now most centres of civilization were close to sea level.

This pronouncement was treated with derision and contumely. Ozone depletion, and the dangers from nuclear power, were the main interest to governments and environmentalists alike. Soon the CFCs reached five parts per billion in the air and ozone holes appeared over the poles. By themselves the ozone holes were of no consequence, since nothing lived at the ice-age polar regions, but their presence was enough to tip the balance in favour of legislation to ban the use of CFCs. Unfortunately it was too late, for the greenhouse balance had also tipped and the planet was now like a boat passing over the edge of a waterfall, moving ever faster towards the heat of the interglacial. The polar ice was already melting and within a few hundred years all of this Atlantean civilization was deep under the ocean. The legend of a flood and of a great empire beneath the ocean persisted. The stories about it were reiterated over the camp fires of the wandering tribes of hunters.

If there is any moral to be drawn from this tale it is that we are very lucky to have chosen to pollute the air now when the planet is least sensitive to perturbation by greenhouse gases. But if you look at the earth, as I do, as a superorganism then we need to make sure that some other surprise may not be waiting to do to us some other unexpected damage—surprise as great as that which confronted those imaginary Atlanteans.

So let me conclude with some further thoughts about the dangerous illusion that we could be stewards of the spaceship earth.

Everyone these days is or aims to be a manager, and this may be why we talk of managing the whole planet. Could we, by some act of common will, change our natures and become proper managers, gentle gardeners, stewards, taking care of all of the natural life of our planet?

I think that we are full of hubris even to ask such a question, or to think of our job description as that of stewards of the earth. Originally, a steward was the keeper of the sty where the pigs lived; this was too lowly for most humans and gentility raised the 'Styward' so that he became a bureaucrat, in charge of men as well as pigs. Do we really want to be the bureaucrats of the earth? Do we want the full responsibility for its care and health? I would sooner expect a goat to succeed as a gardener than expect humans to become stewards of the earth, and there can be no worse fate for people than to be conscripted for such a hopeless task; to be made accountable for the smooth running of the climate, the composition of the oceans, the air, and the soil. Something that, until we began to dismantle creation, was the free gift of Gaia.

I would suggest that our real role as stewards on the earth is more like that of the proud trade union functionary, the shop steward. We are not managers or masters of the earth, we are just shop stewards, workers chosen, because of our intelligence, as representatives for the others, the rest of the life forms of our planet. Our union represents the bacteria, the fungi, and the slime moulds as well as the *nouveau riche* fish, birds, and animals and the landed establishment of noble trees and their lesser plants. Indeed all living things are members of our union, and they are angry at the diabolical liberties taken with their planet and their lives by people. People should be living in union with the other members, not exploiting them and their habitats. When I see the misery we inflict upon them and upon ourselves I have to speak out as a shop steward. I have to warn my fellow humans that they must learn to live with the earth in partnership, otherwise the rest of creation will, as part of Gaia, unconsciously move the earth itself to a new state, one where humans may no longer be welcome.

REFERENCES

Holligan, P. M., Violler, M., Harbour, D. F., Camus, P., and Champagne, P. M. (1983). Satellite and ship studies of Coccolithophore production along a shelf edge. *Nature*, **304**, 339–42.

Lovelock, J. (1979). *Gaia*. Oxford University Press.

Lovelock, J. (1988). *The ages of Gaia*. Oxford University Press.

Schneider, S. H. and Londer, R. S. (1984). *The co-evolution of climate and life*. Club Books, San Francisco.

Vernadsky, V. I. (1986). *The biosphere*. Synergetic Press, London.

Watson, A. J., Robinson, C., Robertson, J. E., Williams, P. J. le B., and Fastian, M. J. R. (1991). Spatial variability in the sink for atmospheric carbon dioxide in the North Atlantic. *Nature*, **350**, 50–3.

6
Monitoring the Ocean
John Woods

Dr John Woods has been Director of Marine and Atmospheric Sciences at the Natural Environment Research Council (NERC) since 1986; he is one of Britain's leading oceanographers. During six years as Senior, then Principal Research Officer with the Meteorological Office, he engaged in pioneering research into turbulence and fronts in the upper ocean. In 1972 he was appointed Professor of Oceanography at the University of Southampton and from 1977 spent nine years as Director of Regional Oceanography at the Kiel Institut für Meereskunde in Germany, where the focus of his research was the warm-water sphere of the North Atlantic. A few days before delivering this Linacre Lecture, John Woods had been asked by the British Government to help co-ordinate the Western scientific response to the release by Iraq of vast quantities of Kuwaiti oil into the Gulf. He is the author of several books and numerous papers on marine and atmospheric physics.

INTRODUCTION

Major advances in oceanography depend on new technology. Modern surveying of ocean currents started 200 years ago when James Rennell used the marine chronometer to map the Aghulas current. Today, scientists at the Natural Environment Research Council's Rennell Centre for Ocean Circulation are using the satellite altimeter to map surface currents in the World Ocean Circulation Experiment (WOCE). Modern derivatives of the Swallow float, which revolutionized understanding of deep ocean currents in the 1960s, will also be used extensively in WOCE. Measuring acoustic travel–time provides a cost-effective method for monitoring the deep eddying motions discovered by Swallow. The NERC *Autosub* project will automate sampling of the deep ocean by means of unmanned submersibles capable of descending to depths of six kilometres along megametre tracks. In order to exploit the flow of data from such instruments oceanographers are developing novel applications of information technology, whereby the observations will be automatically assimilated into mathematical models of the ocean, as in the Fine Resolution Antarctic Model (FRAM) project. Plans are being made to deploy these new techniques in a Global Ocean Observation System (GOOS) for forecasting climate in the twenty-first century.

Thus, the theme of my lecture is the contribution that novel instruments make to the science of the sea. Historically, all major advances in oceanography have followed the introduction of new instruments (I shall give some examples shortly). However, mariners tend to be conservative by nature, and in 1991 marine scientists are still using techniques that have scarcely changed since 1891. Now, we are facing a rapidly rising demand for data which cannot be met by the old manual methods. To satisfy this demand, we shall need a range of new instruments capable of wide deployment around the World Ocean. To take a metaphor from the Gulf War, we need to introduce 'smart weapons' into the oceanographer's armoury. To be realistic, it will take 20 years to introduce such novel technology. But recently there was a call in Parliament for an accelerated programme designed to complete it in ten years (Hansard, 6 February 1991). If that target were accepted by governments in Britain and elsewhere there would be a demand for engineering skills comparable with those employed in landing men on the Moon.

That is a bold vision, which I must justify. My plan is to start with a brief historical review of how our knowledge of ocean circulation has been acquired, culminating in a major advance in the last few months. That breakthrough (and I believe the overworked term is appropriate in this case) opens the way to operational forecasting of the oceans. The scientific issues are whether there will be useful skill in such forecasts and how far they extend into the future. The engineering issue is how to develop a cost-effective system for collecting the data needed to support ocean forecasting. The political challenge is how to achieve international agreement to fund and deploy that observing system, both in space and at sea.

Assuming those challenges can be successfully met, it will be possible to embark on operational forecasting of ocean currents and their influence on climate, fisheries, offshore operations, waste-disposal, and pollution. The potential benefits are vast; greater even than those from weather forecasting. More deaths can be prevented by ocean forecasting than by weather forecasting, and our economic and social well-being are more vulnerable to change in the ocean than in the atmosphere.

On this occasion it is appropriate to focus on the technological issues. My lecture will concentrate on new instruments being used to survey the circulation of the ocean. I shall indicate how they might be deployed to produce the Global Ocean Observing System (GOOS) needed for ocean forecasting. Let us start by going back 200 years to the first scientific measurement of ocean currents.

THE LAST 200 YEARS

Two hundred years ago the invention of the marine chronometer revolutionized measurement of longitude. Captain Cook used one like *Harrison's No. 4* (now in the National Maritime Museum, Greenwich) to survey the Pacific. But it was a major in the East India Company, James Rennell, who first used the new instrument to measure ocean currents. He computed the speed and direction of the surface current from the vector difference between the ship's motion through the water and its change in latitude and longitude (the latter being measured by chronometer). Rennell first mapped the Aghulas current, which brings water from the Indian Ocean into the Atlantic around the Cape of Good Hope. This was one of the most troublesome currents in the world (especially for the East India Company). We shall encounter it again during my lecture. After publishing the chart of the Aghulas current in 1778, Rennell embarked on an unprecedented project: to map the currents of the Atlantic. The result was published posthumously in 1832. In his Atlantic monograph (Rennell 1832), he also revealed that he had already embarked on the yet larger task of mapping the currents of the Pacific. Recently in Southampton, Lord Rennell of Rodd opened the new James Rennell Centre for Ocean Circulation, which has been established to complete the work begun by his famous ancestor.

Rennell was the first to map the *surface* currents of the ocean scientifically. One hundred years later (in 1872) Wyville Thompson on HMS *Challenger* started the task of mapping the currents in depth. The methods used in 1874 were not very different from those used on board HMS *Hecla* in 1974. The ship stops at a *hydrographic station*, where instruments and water bottles are lowered on a 10 kilometre-long wire, to measure temperature and sample water for recovery and analysis in the ship's laboratories or ashore. The procedure is time-consuming and requires great skill on the part of the ship's crew to keep the wire vertical in rough weather. (*Challenger* was equipped with a steam engine to hold station and a donkey engine to pull in the wire.)

Given the trouble it takes to collect hydrographic data, and the precision needed (temperature to 1 millidegree; salinity* to one part per million) it is not surprising that only a handful of laboratories around the world are committed to the task. Nor is it surprising that only a few thousand precision hydrographic stations have been established since the *Challenger* expedition. If the ocean is divided into $5° \times 5°$ squares, many have never been sampled, and many have not been visited since the *Challenger*.

The oceanographic database is, to say the least, sparse. Even in the best observed ocean—the Atlantic—it is difficult to describe the three-dimensional distributions of temperature and salinity (and therefore density, pressure gradient, and current

---

* Salinity is the concentration of dissolved salt, which varies significantly through the ocean. The density of sea-water depends on pressure (i.e. the depth), temperature, and salinity.

velocity) with much confidence. The general features are known, but the description is fuzzy, because the data were collected irregularly over 100 years, during which time the currents have changed with the climate. And in the 1960s it was discovered that the problem was much more complex than we had first thought, when John Swallow found that deep ocean currents do not flow smoothly, but chaotically (Swallow 1971). In effect, the ocean contains its own weather systems, which are much smaller than those in the atmosphere. The transient storms which make up the ocean's weather contain 99 per cent of its kinetic energy; only 1 per cent is attributable to the permanent currents. This means that a very high sampling density is needed to produce statistically significant maps of the permanent currents. Even if the non-synopticity of the data in oceanographic archives is ignored, the total mass of data is quite inadequate to sort out the permanent current *signal* from the weather *noise*.

To resolve this problem, oceanographers have embarked on a massive programme of data collection, which should remedy the situation by the end of the century. Meanwhile, new methods are being developed to extract more information from existing data. This brings me to the first theme of my lecture: the interplay between observations and models. The underlying philosophy is easily stated. It concerns the way in which ocean data are analysed. The issue is how to relate observations to the laws of Nature. Shortly, I shall show how that is done through the use of mathematical models, but first let us recall the methods that have been used during the last hundred years.

CLASSICAL ANALYSIS OF OCEAN OBSERVATIONS

The classical approach to understanding the circulation of the ocean has been to tackle the problem in two stages: first the data were analysed statistically, then the results were diagnosed allowing for the laws of Nature. The values of temperature, salinity, density, chemical concentrations, and current velocity were calculated from the observations, then plotted in vertical sections or in maps. Traditionally they are presented in oceanographic atlases (Stommel and Fieux 1978). Nowadays, the process of plotting can be automated with the help of statistics packages available for computers. Note that in producing oceanographic atlases, the values of temperature, salinity, and so on, are treated as numbers. But that does not do them justice. Actually, they are not merely numbers: they are samples of a dynamical system which obeys the Laws of Nature. Treating the data statistically ignores that fact, with the result that much of the information contained in the observations does not reach oceanographic atlases.

This has long been recognized by oceanographers. The missing ingredient is the Laws of Nature, including those for the dynamics of fluids, and relevant chemistry and biology. The classical solution has been to take account of those laws by diagnosing subjectively the distributions derived statistically from observations. Our understanding of the oceans comes from such diagnosis. Its success depends on the skill of the expert responsible. Its limitations arise from the inevitable limitations of the human mind in coping with the complexity of the ocean. In particular, human diagnosis inevitably tends to linearize the strongly nonlinear interactions that control the marine environment. The challenge is to develop a new method of diagnosis that can cope with the nonlinearities. The answer comes from meteorology.

Increasingly over the last 20 years, meteorologists seeking to understand the circulation of the atmosphere have turned to mathematical models to diagnose their observations. In fact, nowadays, meteorologists concerned with the global circulation seldom if ever examine the raw data (even though they are synoptic) or charts derived from them by statistical methods. Rather, they diagnose distributions generated by mathematical models which have been integrated forward in time for a few hours while assimilating the observations. The model treats the observations not as mere numbers but as samples of variables simulated by a model designed to obey the Laws of Nature. This approach has led to substantial advances in the understanding of atmospheric circulation. Can it help us to understand the ocean?

MATHEMATICAL MODELLING OF THE OCEAN

The fluid motion of the ocean is described by the same dynamical equations as those used in weather forecasting models, in particular the Reynolds equation for the acceleration of currents.* The flow in the ocean closely mimics that of the atmosphere. However, there is one fundamental difference: that of scale. The planetary waves, storms, and fronts which make up the weather inside the ocean are much smaller than their dynamically identical counterparts in the atmosphere (Fig. 6.1). A single storm over the Atlantic may cover 1000 storms inside the ocean. The Rossby radius of deformation, which determines the dimensions of storms, lies in the low tens of kilometres in the ocean and around 1000 kilometres in the atmosphere.† The Rossby radius also determines the widths of permanent ocean currents like the Gulf Stream.

If ocean currents are to be accurately simulated by a mathematical model of the ocean, the model must be integrated using a grid-spacing not much larger than 10 km. Global coverage at 10 km resolution requires five million grid points for every layer; a total of 150 million points for a model with 30 levels.‡

---

* Reynolds equation is the fluid flow equation governing the behaviour of quantities (e.g. velocity) resolved by the grid of a numerical model of fluid motion. It includes terms designed to deal statistically with smaller scale motions which exist in Nature, but which cannot be resolved by the model.
† The Rossby radius of deformation is the horizontal scale at which gross rotation effects become as significant as buoyancy effects in the equations governing fluid dynamics. Storms are rotational in nature, and therefore have a scale governed by the Rossby radius.
‡ For comparison, it is possible to forecast the weather with a model that has only a quarter of a million grid points, in ten layers.

Fig. 6.1 Atmospheric storms cover up to 1000 storms inside the ocean. (a) Satellite image of clouds marking a storm in the atmosphere (*Source*: NASA.). (b) Satellite infrared image of sea surface temperature patterns marking a storm in the ocean. (Reproduced with permission from Dundee Satellite Station.) The scales of the two storms are 1000 km for the atmosphere (a) and 50 km for the ocean (b).

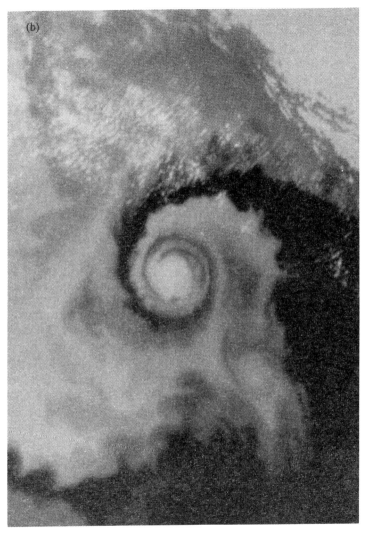

The central technical problem of ocean modelling is to achieve the fine resolution needed to simulate ocean currents. Later, we shall see that even finer resolution is needed to simulate plankton growth which depends on mesoscale upwelling with horizontal dimensions of one to ten kilometres.* This is a daunting specification that is only becoming feasible in the 1990s thanks to the increasing power of supercomputers. To illustrate the state of the art in 1991, I shall now present some results from the British FRAM project, which is in the vanguard internationally.

THE FINE RESOLUTION ANTARCTIC MODEL (FRAM)

The aim of the FRAM project is to simulate the currents in the Southern Ocean using a model integrated with a 27 km grid and 32 layers, and initialized by the hydrographic data from the World Data Centre. Those data were accumulated irregularly over the last 100 years. They are certainly not synoptic. We know that the climate of the ocean varies over a broad spectrum spanning years, decades, and centuries. So treating the data statistically as though they were synoptic leads to internal inconsistencies, and fuzziness in the computed distributions and currents.

The first task of the model was to generate clear synoptic distributions consistent with the Laws of Nature and with the observations. To achieve that the archive data were assimilated into the model as it was integrated forward in time over a simulated period of six years (Fig. 6.2). During that period of data assimilation, the model wrestled with the inconsistencies of non-synoptic data and the Laws of Nature (as described by the nonlinear laws of motion). Twenty million variables were computed at every step in the integration. No human mind could have performed such elaborate, nonlinear diagnosis of the data. Yet there was no little human judgement involved in designing the method of data assimilation. In fact there is scope for more

---

* Mesoscale upwelling is the localized (relatively) rapid vertical transport of cold, nutrient-rich ocean water to the surface, caused by confluence between ocean eddies. The nutrients in the upwelling water are made available to surface plankton at such a rate that massive plankton growth rates become possible. The importance of plankton for the greenhouse effect is described by James Lovelock in his Linacre Lecture (see Ch. 5).

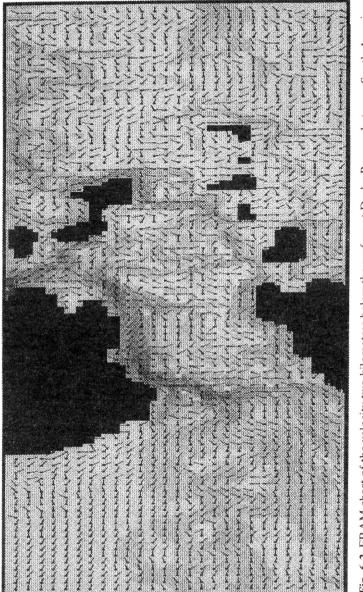

Fig. 6.2 FRAM chart of the velocity two kilometres below the surface at Drake Passage between South America and the Antarctic peninsula. The speed is indicated by the shading, the direction by the pointers.

research into the technique, exploiting the mathematics of chaos theory.

The FRAM integration has been used to produce the world's first oceanographic atlas based on data assimilation into a high-resolution mathematical model. All previous atlases have been based on the classical method in which data were treated as numbers rather than samples of a dynamical system. It has cost over two million pounds to produce the FRAM Atlas (Webb *et al.* 1991). Was it worth it? What have we learnt? The answer to the first question is a resounding 'yes' because it gives for the first time a realistic description of the Southern Ocean circulation. And to the second question, we can reply that the way in which we perceive ocean currents has been transformed by the FRAM Atlas, with important implications for our understanding of the ocean, and its role in climate.

The FRAM discovery centres on the interaction of ocean currents with bottom topography (Fig. 6.3). The deep ocean floor has an average depth of 4 km but it is not flat. The Challenger Deep is 11 km. And the world's most extensive mountain ranges rise up from the ocean floor, reaching on average to within two kilometres of the surface, a dynamical limit controlled by deformation of the mantle. Exceptions occur in regions of intense volcanic activity, where mountains may rise through the surface to give mid-ocean islands, such as Iceland, the Azores, and Hawaii, but these are rare. The submerged mountain ranges control the circulation of dense water which fills the deep ocean, but it had normally been assumed they do not significantly influence the strong surface currents like the Gulf Stream. Classical analysis of hydrographic data led us to believe that the kinetic energy of such currents is concentrated into the top kilometre or so of the ocean. That diagnosis led to an important assumption in classical dynamical oceanography, namely that there was a *level of no motion* at about one kilometre below the surface. The vertical profile of current velocity was computed by integrating the shear (derived from hydrographic data) upwards from such a level of no motion. The FRAM Atlas has destroyed that paradigm.

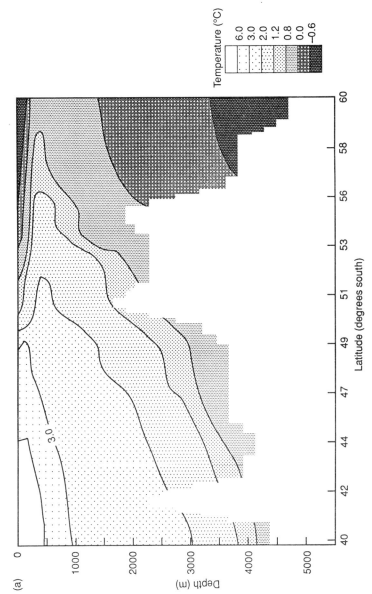

Fig. 6.3 Vertical sections of temperature (a), salinity (b), and velocity (c) across the Antarctic Circumpolar current, showing the interaction with submerged mountain ranges discovered by the FRAM project.

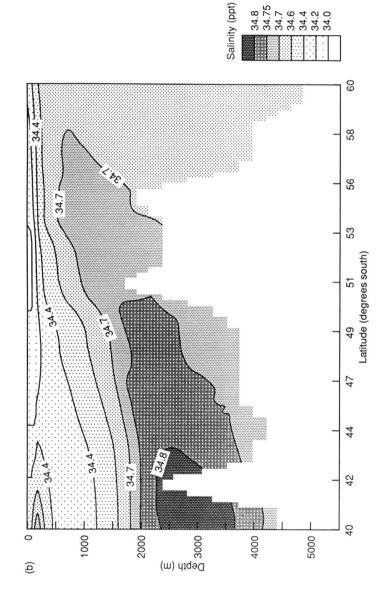

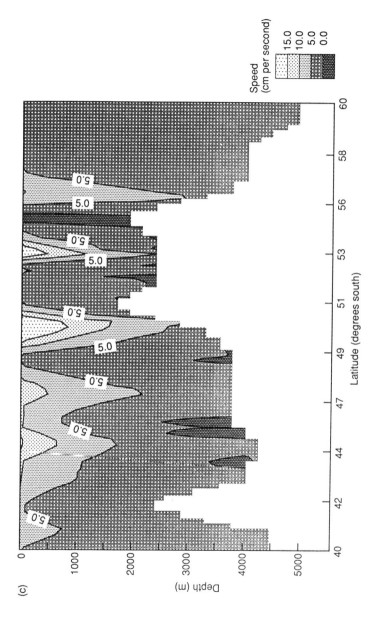

According to FRAM, permanent currents are not concentrated in the top kilometre, but reach right down to the submerged mountain ranges, which they cross in narrow jets lying on the flanks of mountain passes. This new paradigm for ocean currents echoes the few measurements that we have of the structure of ocean currents, and it is consistent with the dynamics of ocean storms, which are known to interact strongly with bottom topography. It provides support for conjectures that the structure of the large-scale permanent currents which make up the general circulation of the ocean is largely controlled by the energetic, but transient, weather systems of the ocean.

The FRAM paradigm is important for understanding how the circulation changes with the climate because it leads us to abandon the notion based on classical diagnosis that the currents can shift their tracks smoothly, unimpeded by the submerged mountain ranges beneath them. Now we recognize that the currents are tightly bound to those mountain ranges and, if the winds driving them shift, the currents must adjust in steps from one mountain pass to the next, or change the distribution between multiple jets flowing over a set of neighbouring passes.

This interaction between currents and topography can only be simulated by a model that has sufficiently fine resolution to capture the processes which effect it. Those processes have horizontal scales close to the Rossby radius, that is, a few tens of kilometres. There is no known way to parametrize the processes in a model with coarse resolution.* We conclude that models designed to forecast climate may have to employ a grid spacing of about ten kilometres. At present climate models have grid spacing of about 150 km, which is too coarse to resolve the Gulf Stream, let alone its internal structure.

---

* To parametrize a process is to treat it by terms in the model equations based on a statistical treatment of its effect. All unresolved processes must be parametrized in models. The most familiar example is parametrization of turbulent transport by means of eddy diffusivity, which does not work for the interaction of currents with submerged mountain ranges.

How realistic are the simulations?

The FRAM group is now comparing the simulated descriptions with observations. Their ability to do this is limited by a lack of suitable data. Some questions can be answered by archive data: for example, earlier models have not succeeded in getting major currents like the Gulf Stream to follow the observed mean track across the oceans, and in particular to leave the coast in the right place. The currents simulated by FRAM do much better: for example, the convergence of the Falklands current and the Brazil current is described well, as is the pulsating flow of the Aghulas current. The availability of new types of observations is helping this diagnosis. In particular, we can compare the surface dynamic topography mapped by the Geosat altimeter with that generated by FRAM. The results are most encouraging. So, too, are the tracks of eddies moving from the Indian Ocean into the Atlantic.

Heat circulation

Diagnosis of the FRAM output is also throwing up some unexpected results which we shall not be able to check until we have new data. For example, it is possible to calculate the heat carried by the currents. (The circulation of heat by currents in the ocean corrects the local imbalance between solar heating and heat loss to the atmosphere). We have known for some time that the ocean currents circulate heat globally at the same rate as the atmosphere. One of the major features of the global heat flux is a northward flow in the Atlantic which emerges to power the westerly winds over the North Atlantic. According to classical analysis, that northward heat flux of about one billion megawatts is supplied equally from the Pacific and Indian oceans. In the distributions generated by FRAM, the flow comes overwhelmingly from the Indian Ocean, via the notoriously variable Aghulas current. James Rennell (1832) was the first to suggest that fluctuations in the climate of Britain depend on those in the Gulf Stream: perhaps they also depend on fluctuations in the

Aghulas current. If correct, the FRAM discovery will influence our thinking about how to predict climate. But before addressing that issue let me briefly refer to another development: the introduction of biology into ocean modelling.

Plankton ecology

In recent years a set of equations has been developed which decribes the principal processes in plankton ecology, such as photosynthesis, grazing, predation by carnivores, and the vertical motion of plankters with and through the water. Methods are now well established for integrating these equations in one-dimensional models which show the interaction between the plankton and the turbulence, light, and chemical variations in the vertical at one location. Such simplified models capture many of the features of the seasonal bloom of plants and their consumption by herbivores. The models reveal how plant growth consumes carbon dioxide dissolved in the seawater, causing more to flow in from the atmosphere: this is a major factor in the greenhouse effect. In order to compute the regional variations of these processes it will be necessary to take account of the ocean circulation. A first step is being taken by adding biological equations to models such as FRAM. The new data needed to test these models of plankton ecology come from novel instruments in space and in the ocean, as we shall see below.

PREDICTABILITY

Now I come to the key scientific issue that I mentioned earlier: to what extent is the ocean predictable? Ocean circulation is chaotic, like the atmosphere's. Their dynamics are described by the same nonlinear equations, so we can expect them to have similar properties regarding predictability. Of course, there are significant differences in the space and time-scales of dynamically equivalent phenomena in the ocean and atmosphere: in terms of area, as we have seen, storms are 1000 times larger in the atmosphere than in the ocean, but ocean storms persist for ten times as long. The issue of predictability is now well understood in meteorology, thanks to the pioneering work of Lorenz and others, who have established the rate of error growth in weather forecasting models, and the ultimate limits to useful prediction.* Their theories have been verified by operational forecasting centres, where, thanks to progress in modelling techniques, the limits to predictability are dependent increasingly on the initial error in the observations used to start the integration. Today, meteorologists are all too aware of their dependence on the international system, called World Weather Watch (WWW), used to collect observations for weather forecasting. Efforts are being made to improve WWW, but even if they succeed, it will never be possible to predict the weather more than a week or so ahead, the limit determined by chaos theory. The atmosphere has no 'memory' beyond one month.

Gulf oil pollution

What lessons can ocean forecasters learn from the experience of meteorologists in this matter of predictability? Some ocean forecasts are so dependent on local weather that they suffer exactly the same limits of predictability as weather forecasts. Examples include the prediction of sea-state and storm surges. These came together dramatically during the Gulf War in efforts to predict what would happen to the oil that Iraq released from Kuwait into the Gulf. We were asked by the Government to produce (in one week!) a mathematical model to simulate the changing distribution and properties of the pollution in the Gulf. The model (Fig. 6.4) describes in detail how the under-

---

* The limits to predictability of the climate system are not yet known either theoretically or practically, but climatologists assume that the nature of climate predictability will be similar to that for weather prediction (because similar non-linear equations are involved), with the difference being that, for climate, the dynamics are controlled by the coupled ocean–atmosphere system.

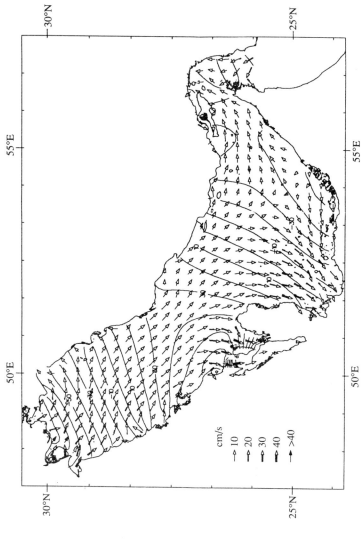

Fig. 6.4 A model of the circulation in the Gulf used to predict oil pollution during the 1991 Gulf War.

water plume of toxic water progressed south through the coral reefs and breeding grounds of Gulf marine fauna. This hindcast was invaluable for analysis of ecological impact after the war. Can the model also be used in forecast mode? Yes, but the currents depend on wind as well as tide, so the accuracy of such predictions is limited to that of weather forecasting, namely a few days. Forecasting further into the future must be based on climatological-mean winds. Does that limit apply to all aspects of ocean forecasting? We have been led to expect that the ocean has a longer memory than the atmosphere. What evidence is there for that?

The El Niño–Southern Oscillation

The first attempt to measure the predictability of ocean models has come from work on the El Niño–Southern Oscillation (ENSO). ENSO events occur every five years or so (the time taken by Rossby waves to cross the Pacific). They involve large-scale temperature anomalies in the tropical ocean, which induce atmospheric disturbances that affect the climate all round the Pacific Basin. Models designed to predict ENSO events are currently based on a model of the upper Tropical Ocean coupled to the Global Atmosphere (TOGA). It has recently been demonstrated that such models have useful predictability for several months ahead, provided they are supplied with an accurate description of the initial state of the ocean. This predictability depends on the fact that the currents and planetary waves in the upper 100 metres of the tropical Pacific Ocean have a memory that greatly exceeds that of the atmosphere. That memory is retained despite interaction with the atmosphere: in fact there is some evidence of positive feedback in the ocean–atmosphere coupling which sustains the memory of ENSO events. Although it has not yet proved possible to predict an event, that seems to be a real possibility. Today the limiting scientific problem is the accurate simulation of energy and water fluxes between the ocean and atmosphere. Once that is resolved, the limiting factor will be the accuracy of the initial description. As in weather forecasting, the magnitude of the errors in the initial observations will determine the skill in the forecast.

The greenhouse effect

ENSO events are well documented and they produce great economic and social stress, so forecasting them will be very worthwhile. But much larger problems confront us as the result of the greenhouse effect, which is expected to change the climate of the earth during the next century faster than at any time since civilization began (IPCC 1990). One of the most serious consequences of global warming due to the greenhouse effect will be a rise in the sea level during the next century. Attempts are now being made to construct models of the climate system that can be used to forecast the way in which the climate will change during the next century. If such models are to have any predictive skill, it can only be because the climate system has a memory extending over decades. We have seen that the memory of the atmosphere is less than one month. The land has a longer memory, but it is disturbed by man's use of land and water. The polar ice-caps have a very long memory, but they do not change sufficiently over decades to form the basis for climate forecasting. The only component of the climate system with the potential for a long memory is the ocean.

There is some evidence of systematic variation inside the ocean with time-scales of decades, and we know it influences the climate, but it is not yet clear whether it can in principle be predicted. The first goal of the World Climate Research Programme is to discover 'to what extent climate is predictable'. The scientific problem has two components. The first is to discover whether aspects of the ocean circulation—in particular the large-scale fluxes of heat, water, and chemicals carried by ocean currents—have a memory of decades that can be exploited in climate models. The second is to discover whether it is possible to make models of the atmosphere capable of exploiting that memory of the ocean, by accurately simulating: (a) the energy and water fluxes from the sea into the air, (b) clouds and their influence on radiation and precipitation, and (c) interaction with the land surface.

THE WORLD OCEAN CIRCULATION EXPERIMENT

These scientific issues are being addressed by the World Climate Research Programme (WCRP) in a set of major experiments. The World Ocean Circulation Experiment (WOCE) is concerned with the issue of decadal predictability of the ocean. It started in 1990 and will continue at sea until 1997 (Fig. 6.5). The goal is to make the first ever survey of the global circulation of the ocean from top to bottom, sufficient to describe the circulation of heat and water to the accuracy specified by the WCRP. That specification translates into calculating the divergence of the ocean heat flux to $\pm 10\,\text{W/m}^2$, averaged over ten degrees of latitude in each ocean basin. That is a daunting task and has led to an experimental design which stretches contemporary observing capability to the limit. The armoury of instruments to be deployed in WOCE far exceeds anything attempted before by oceanographers. To quote one statistic: WOCE requires our best research ships to make 24 000 hydrographic stations along carefully designed trans-ocean lines. For comparison, the existing database used to determine ocean circulation contains useful hydrographic profiles at only 8000 stations, accumulated over 100 years. This great survey will be seen historically as a turning point in oceanography. It will open the way to operational forecasting of the ocean circulation.

OPERATIONAL FORECASTING OF OCEAN CIRCULATION

Let me recapitulate. The memory of the atmosphere is less than one month, but it is believed that the memory of the ocean is much longer. It has recently been demonstrated that the error

146 *Monitoring the ocean*

Fig. 6.5 The World Ocean Circulation Experiment (WOCE) will involve hydrographic stations at 50 km intervals along the trans-ocean lines shown here.

growth in models of ENSO events are consistent with a memory of several months. It has not yet been possible to perform modelling studies to discover whether there is useful decadal memory in the large-scale circulation of heat and water, but there is indirect evidence in the form of decadal anomalies in temperature and salinity.

We expect to find from research now going on that the ocean circulation does have an intrinsic memory of several decades. The WOCE data-set will provide one starting point for climate forecasting. Verification of the forecast will depend on collecting further data-sets in the future. Each of those new surveys of the state of the ocean circulation can in turn be used to initialize further independent forecasts extending up to a century ahead. Once data collection and model integration become routine we shall have an operational system for ocean forecasting, equivalent to the system running today for operational weather forecasting.

Computers

We are proceeding on the basis that the ocean circulation does have decadal memory. (If it does not there is no possibility of forecasting the climatic consequences of the greenhouse effect.) Modelling of the global ocean circulation is advancing as computer power permits. It is estimated that a single integration of a global ocean circulation model will involve a billion billion floating point calculations—or one Exaflop. That will take an hour on a Teraflops-terabyte machine, which should be available within ten years.* So, by the time the WOCE data-set is processed, we should have computers capable of running models that can assimilate the data efficiently. The memory of the ocean can then be computed by a series of integrations of a coupled ocean–atmosphere model, each starting with initial conditions that vary differently from the WOCE data-set. Once the potential for prediction is established we can proceed to operational forecasting using the same models. Each new forecast will require a fresh data-set describing the state of the ocean at the start of the integration. Experience in weather forecasting has shown that the quality of the observations plays a key role in determining how closely forecasting can approximate to the theoretical limits of predictability. Plans are already being prepared for an observing system for climate forecasting.

* Experimental machines now being developed encourage the view that computers will continue to become ten times faster every five to six years. On that assumption, a petaflops machine (i.e. one capable of a billion million floating point operations every second) should become available by 2015, making it possible to simulate ocean circulation and plankton globally.

The Global Ocean Observing System

The experience gained in designing WOCE provides a pretty clear idea of what it will take to make an operational system for ocean circulation forecasting. The major task will be to create a Global Ocean Observing System (GOOS) designed to provide a flow of data at monthly intervals for initialization of ENSO forecasts and at annual intervals to initialize forecasts of decadal change. Later this year governments will be asked to establish an International Planning Office for the Global Climate Observing System, which includes GOOS.* Preliminary estimates suggest that the cost of running a GOOS are likely to be comparable with those of World Weather Watch, that is about $2 billion per year.

The design of GOOS must serve the rather different needs of forecasting on inter-annual and decadal time-scales. It turns out that a single observing system can meet the requirements of both types of forecasting. That is because the fluctuations in the ocean–atmosphere heat flux are much greater in ENSO events ($100\,W/m^2$) than in the greenhouse effect ($10\,W/m^2$). The density of observations needed to describe the ocean to within ±20 per cent of those signals can be achieved by tapping the data collection cycle every month for ENSO forecasting but accumulating a year's worth of monthly samples for decadal forecasting. We shall see how this works out in practice shortly. First we need to review the armoury of instruments that will be available to the designer of GOOS.

TOOLS FOR OBSERVING THE OCEAN

Not all instruments used by oceanographers are suitable for long-term, world-wide deployment as part of a global ocean observing system; I shall concentrate on those that are. In general we can divide them into instruments deployed in space and instruments deployed inside the ocean. A preliminary

---

* Subsequently approved for establishment in early 1992.

analysis suggests that the cost of GOOS will depend about equally on those two categories.

Satellites

The first instruments to monitor the ocean from space were scanning radiometers in the visible, infra-red, and microwave bands. These have given useful images of the distributions of plankton, temperature, and sea ice respectively (Fig. 6.6). Their main value has been in revealing patterns on scales dominated by the eddies and fronts which make up the weather inside the ocean. The solar energy flux entering the ocean has been estimated to $\pm 10\,\text{W/m}^2$ from statistical analysis of the gaps between clouds detected by visible and infra-red radiometers. Global distributions have been produced by combining images collected over a month. However the resulting maps of surface temperature have a typical error of $\pm 2\,\text{K}$ which is too large for GOOS. A new instrument, the 'along-track scanning radiometer' (designed at Oxford University and the Rutherford Appleton Laboratory) is expected to get the error down to a few tenths of a degree, which *will* be useful for GOOS.

Those instruments are all passive radiometers. The era of active radars began in 1978 with the launch of the Seasat mission. They will feature strongly in the first European Ocean Observing Satellite (ERS-1) due to be launched in May 1991.* The radars include an altimeter, which can measure the distance between the satellite and the sea surface to an accuracy of 2 cm (over one million metres!), and a scatterometer, which maps the strength of the return signal. The altimeter is used to measure the horizontal variation of the dynamic height of the ocean surface: the pressure head that drives ocean currents. The altimeter data are used to compute the velocity of the surface currents, the energy of ocean storms, the tides, wind waves, and the upper ocean heat content. The scatterometer data provide an estimate of the surface wind, the rate at which momentum is entering the ocean, the wind-driven surface flow,

---

* Successfully landed in July 1991 and now operational.

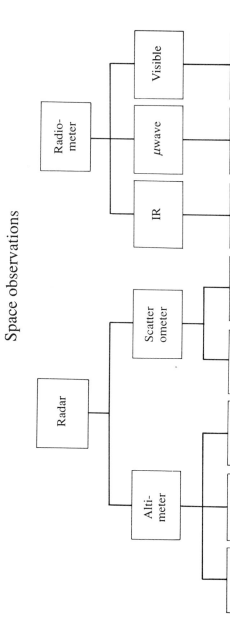

Fig. 6.6 Instruments mounted on artificial satellites can monitor many key aspects of the ocean.

known as Ekman transport, and the convergence of that transport which powers deep ocean currents.

These are unique measurements that cannot be obtained systematically all round the globe in any other way. However they all refer to the surface of the ocean. To achieve the GOOS objective of measuring the heat flux carried by ocean currents, it is necessary to support the global measurements from space with measurements inside the ocean.

Instruments inside the ocean

Instruments in the ocean are deployed in three ways: fixed to the sea floor, drifting freely with the currents, or motoring around the ocean.

Fixed instruments

A global network of 300 coastal sea-level gauges (GLOSS) will soon be operational to monitor heat content, tides, and surface currents. They will be supplemented by a smaller array of pressure gauges and upward-looking echo-sounders on the deep ocean floor, to monitor fluctuations in the pressure gradient across major currents. Together, the coastal and ocean floor gauges will provide calibration data for the satellite altimeter.

Moored current meters will be deployed in key areas, such as choke points at narrow sills between ocean basins and in regions of intense eddy energy. The standard current meter has a mechanical rotor. But acoustic remote sensing, based on doppler shift, is now coming in, with the leading instrument designed in Britain at the Proudman Oceanographic Laboratory. Another British invention, called ATTOM, monitors variation in temperature and velocity by changes in the travel time of acoustic pulses. The Americans have extended this technique to acoustic tomography in which the three-dimensional structure of temperature and velocity can be reconstructed from the travel times of pulses radiated between hydrophones in a ring. Acoustic travel time has also recently been tested successfully over global ranges from Heard Island in the Southern Indian Ocean to measure the mean temperature of the ocean to a fraction of a millidegree.

Drifting instruments

In the very deep ocean, currents often have a mean speed that is so slow that the associated pressure gradient is smaller than the resolution of satellite altimetry (say, 2 cm per megametre). We measure such deep currents by tracking the drift of floats balanced to lie on a chosen pressure surface. The floats are either tracked acoustically from a set of receivers moored around the chosen ocean basin, or by satellite when the floats briefly surface according to a programmed schedule. John Swallow was first to design a float to lie on a pressure surface (the trick is to make the compressibility of the float less than that of seawater). He discovered that the deep ocean is filled with energetic eddy motion: a discovery that changed our paradigm for ocean circulation.

In order to map the deep velocity field, it is necessary to seed the ocean with sufficient floats to ensure that a statistically significant number pass through every $10° \times 10°$ square every month. A very large number of floats is needed if the aim is to map the weak mean flow in the presence of the energetic eddies; a signal-to-noise ratio in the order of 1 per cent. The WOCE design calls for several thousand floats.

Drifters floating on the surface can also be used to track currents. They are particularly useful in the tropics where the wind waves are relatively weak. Several hundred floats are deployed in the Pacific Ocean for the TOGA study. Surface drifters are easy to track by satellite using the Argos system which can also receive and transmit to the user observations of the surface temperature and atmospheric pressure.

Mobile instruments

The third category of ocean sampling is based on platforms that move through the water. They include research ships which move from station to station along prescribed tracks, taking several hours to lower instruments at one metre per second to the ocean floor 5 km down. Temperature and electrical conductivity are measured continuously as the instrument descends, spot values are made of temperature with mercury reversing thermometers, and water samples are collected by Nansen bottles for later measurement of salinity and chemical composition. For dynamical oceanography, the distribution of seawater density must be calculated to four significant figures from the temperature (measured to one millidegree) and salinity (measured to one part per million). Achieving these accuracies has not proved easy and only a handful of laboratories in the world regularly achieve it. Only a few thousand of the hydrographic profiles in our archives meet the specification: most were made for other purposes and cannot be used to compute ocean currents. Stringent controls have been designed into the plan for WOCE to ensure that the specification is met.

Precision hydrography from ships, which has barely changed in the last hundred years, remains the only method for determining the vertical variation of ocean currents (by using the horizontal density gradient in the geostrophic balance equation). Oceanographers compute the velocity profile by integrating the shear from a reference level at which the velocity is known. The reference level velocity is determined either at the sea surface from altimetric measurement of the horizontal pressure gradient, or deep in the ocean from the statistics of drifter tracks. Thus the precise measurement of temperature and salinity profiles at stations along lines across the ocean lies at the heart of ocean circulation monitoring. It is a prerequisite for calculating the heat flux carried by ocean currents.

Most of the ship time in WOCE will be devoted to making some 24 000 precision hydrographic profiles. The enormous task will take the equivalent of 15 years on one of the small number of first-class oceanographic ships in the world fleet. Looking to the future, when ocean circulation is to be monitored routinely for climate prediction, there will be a demand for the equivalent of a WOCE every month.* It is unrealistic

---

* If the forecasts are to simulate the climatic change caused by the greenhouse effect, the monthly equivalent of a WOCE, to measure heat flux divergence to a few W/m^2, will be required.

to expect that sufficient ships and people can be found to perform such a task every month, so some new, cost-effective method must be found for *operational* precision hydrography. We believe the answer lies in the use of autonomous, unmanned submersibles, 'autosubs' for short. The NERC is now well on the way to producing a prototype of the Dolphin Autosub which will test whether this is feasible.

Autosubs

The NERC Dolphin Autosub will be a seven-metre-long vehicle weighing five tons with a range of 6000 km and a depth capability of 6000 metres. The initial instrument payload will include a CTD for precision hydrography; later versions will carry chemical and biological samplers. The vehicle will be powered by an oxygen-hydrogen fuel cell driving a novel electric motor, which is now undergoing tests at the Deacon Laboratory of the NERC's Institute of Oceanographic Sciences. To achieve trans-ocean range the vehicle has a low-drag outer form designed by supercomputer and due to be tested at sea in 1992. The pressure container must be lightweight and have the strength to resist the weight of 6 km of water and low compressibility: no metal meets these requirements, so it is planned to use composite materials based on carbon and other fibres in epoxy resin.

Data will be returned by satellite, either when the autosub occasionally surfaces, or when it releases an expendable data capsule to float on the surface. Prototype data capsules have already been tested successfully at the Proudman Oceanographic Laboratory. The position of the vehicle can be fixed by satellite when it surfaces. Trials by a team at the NERC Sea Mammal Research Unit working with the British Antarctic Survey have shown that instrument packs on elephant seals can be used to fix the position of the 'bio-autosub', and data have been transmitted over the Argos system during brief periods when the seal is on the surface during rapid megametre transits.

These and other design features are now being researched in university, government, and industrial laboratories in Britain and other countries. The aim is to test the subsystems in the next two years and make a generic prototype for trials in 1994. If the programme goes as planned, we could have a 'proof-of-concept' science mission in 1995. The cost of the project through to an operational Dolphin Autosub is estimated to be £20 million at 1991 prices.

A number of science projects are being planned for *Dolphin*. The primary mission is to make trans-ocean precision hydrography sections. Later, we hope to develop a version that can do the same under the polar ice. A user community is already meeting annually to design these and other experiments. We believe that in the next century, autosubs will take over routine surveying, leaving research ships to operate in their intended role, as floating laboratories. The autosub method can be scaled up as needed to meet the growing demand for data in a cost-effective way. It is estimated that a precision hydrographic section can be made by autosub for about one-tenth of the cost of doing it in the traditional way, by ship. We are approaching the end of the heroic age of dynamic oceanography. WOCE may well be the last major survey of ocean circulation to be undertaken by research ship.

Operation Vivaldi

The first scientific trial of autosubs is scheduled to take place in the North-East Atlantic as phase two of Operation Vivaldi, a regular survey of the seasonal boundary layer of the ocean. Initially the survey will be carried out by a ship towing an undulating vehicle, *Sea Soar*, which carries an instrument payload similar to that planned for *Dolphin*. The project is designed as a UK contribution to the Gyre Dynamics core project of WOCE. It will also provide the first example of one kind of data that will be collected routinely in the Global Ocean Observing System. Experience gained in collecting, processing, and assimilating the data into a purpose-built model, the Atlantic Isopycnic Model (AIM), will ensure that we know how to handle the data-flow from the autosub when it replaces Sea Soar in the mid-1990s.

CONCLUSION

Recent advances in modelling ocean circulation offer the promise of operational forecasting later in the 1990s when more powerful computers and new observing systems are available. The greatly increased demand for high quality observations will only be met by introducing unmanned instruments in space and inside the ocean. The NERC Autosub project is designed to automate precision hydrography, which will continue to be the key to ocean circulation monitoring, because it alone can reveal the three-dimensional structure of ocean currents. The analysis of hydrographic and other data will increasingly depend on data assimilation into mathematical models integrated with grid spacing small enough to resolve the structure of ocean currents and the processes by which they interact with submerged mountain ranges. Novel instrument technology and information technology are being combined to revolutionize oceanography by the end of the century.

REFERENCES

Hansard (1991). House of Commons debate on the funding of science, 6 February 1991.

IPCC (1990). *Climate change: the IPCC scientific assessment.* Cambridge University Press.

Rennell, J. (1832). *An investigation of the currents of the Atlantic Ocean and those which prevail between the Indian Ocean and the Atlantic.* Rennell Collection, Royal Geographical Society, London.

Stommel, H. and Fieux, M. (1978). *Oceanographic atlases: a guide to their geographic coverage and contents.* Woods Hole Press.

Swallow, J. C. (1971). The Aries current measurements in the western North Atlantic, *Philosophical Transactions of the Royal Society of London* A, **270,** 451–60.

Webb, D. J., Killworth, P. D., Coward, A. C., and Thomson, S. (1991). *FRAM atlas of the Southern Ocean.* NERC, Swindon.

7
The dilemma of the Amazon rain forests: biological reserve or exploitable resource?
Ghillean T. Prance

Professor Ghillean (Iain) Prance was appointed Director of the Royal Botanic Gardens, Kew in 1988, after spending 25 years in the United States. He is one of the world's foremost experts on the Amazon rainforests. After graduation and the award of a doctorate at Oxford University, Iain Prance went to New York on what was intended to be a temporary appointment at the New York Botanical Gardens; in the event, he stayed on to become, in 1968, Curator of Amazonian Botany. He was subsequently appointed Director of Research at the New York Botanical Gardens and then Director of its Institute of Economic Botany. He held Visiting Professorships at the City University of New York and at Yale University. He was Leader of the United States Amazonian Exploration Program for 20 years and visited the region on numerous occasions. He is the author of several books on botany and Amazonia.

Since its discovery by the western world, the Amazon region has been the topic of many stories, exaggerations, and myths that have led to a misunderstanding and misuse of the area. This began with legends about fierce women warriors who gave their name to the region, and of El Dorado, the legendary city of gold, and it has been continued until the present day by exotic tales from adventurous journalists who penetrate to remote parts of the region and by well-meaning environmental activists who vastly overestimate the rate of destruction of the forests. This has always drawn world-wide attention to the region and has been helpful in some respects and harmful in others. It is probably as a result of the work of the media that the rate of deforestation in Amazonia has been considerably decreased during the past two years and that there is some hope of developing a more rational and long-term sustainable utilization plan for the region. But this is unlikely to happen if myths and half-truths are perpetuated. I want to examine here some of the misunderstandings about the region and introduce some of the factors that are vital for the sustainable development of Amazonia.

HISTORICAL BACKGROUND

The real opening up of the Amazon region for development began in the 1960s when Brazil and Peru both began to build major highways into and across their Amazon territories. This was done during the time of military governments, to occupy the land and ensure that they maintained their sovereignty over it. This was the culmination of what, in Brazil particularly, was an apparent paranoia about the ownership of Amazonia. However, when one reviews the history of the region, the territorial disputes that have occurred, and the statements that have been made by various foreign politicians and scientists, it is hardly surprising that foreign interest in the Amazon was and still is questioned. The early history of the regions included territorial disputes between the Spanish, Portuguese, French, British, Belgians, and Dutch. It was the Portuguese who obtained legal possession of most of the Amazon region officially in 1750 by the Treaty of Madrid, but this sovereignty was often to be challenged in the future as just a few notable examples will show.

The French scientist Charles Marie de la Condamine travelled through Amazonia in 1745 to establish the line of the Equator. However, in his map of Amazonia he included the entire Brazilian Territory of Amapá within French Guiana, leading to considerable future disputes about the ownership of the territory. In 1832 a company was formed in London for the purpose of settling the Amazon with English, Scottish, and Irish colonists and it took considerable persuasion from the Brazilians to show that Amazonia was not an abandoned and unoccupied territory. In 1835 the *Cabano* war began as an internal revolution in the Brazilian provinces of Pará and Maranhão. Foreign powers were soon to take advantage of this bloody civil war. The French invaded Amapá and claimed La Condamine's boundaries, which gave them possession of the northern bank of the Amazon river. The British Atlantic Naval Squadron was sent to Belém to help quell the revolution and proposed that the rebels separated from Brazil under British protection. This was followed by an American offer to supply the rebels and establish an American protectorate. After this, a proposal was made by Matthew Fontain Maury, Superintendant of the Natural Observatory in Washington DC, to create new status for the Southern States of the Amazon. He argued that the Amazon river could be considered a natural extension of the Mississippi! Maury's brother-in-law, Lt. William Lewis Herndon, was sent on an exploratory expedition to the Amazon. Maury wrote to him: 'Shall that area be peopled with an imbecile and indolent people or by a go-ahead race that has energy and enterprise equal to subdue the forest and develop and bring forth the vast resources that lie hidden there?' He instructed Herndon: 'I care not what may be the motive which prompts the Gov. to send you there. Your going is to be the first link of that chain which is to end in the establishment of the Amazonian Republic. Herndon's book (Herndon and Lardner 1854) *Exploration of the Valley of the Amazon* (published by Maury) is a direct attack on Brazilian sovereignty and it is hardly surprising that it caused both fear and indignation in Brazil.

Another visitor to the Amazon was the famous Swiss-American scientist Professor Lewis Agassiz, accompanied by his wife Elizabeth, who wrote in their jointly published book *A Journey in Brazil*: 'The time when the banks of the Amazon will teem with a population more active that it has yet seen—when all civilized nations will share its wealth, when the twin continents will shake hands and the Americans from the North come to help the Americans from the South in developing its resources ...' (1969, p. 257). Professor and Mrs Agassiz endorsed the

North American desire to take over the Amazon Valley. In the 1890s, King Leopold II of Belgium expressed concern about a new French occupation of Amapá and offered to create a free state under Belgian protection. This was an attempt to obtain an equivalent of the Belgian Congo in the New World. During the rubber boom, the British offered to set up a gigantic rubber plantation in the Amazon in return for many conditions that would have reduced Brazilian control of the area. The Brazilian refusal of this offer led to a switch of interest to Asia and the transfer of rubber seeds to Asia, via the Royal Botanic Gardens in Kew. In 1890 the Americas attempted to take over Acre Territory in the western Amazon through the formation of a company, 'The Bolivian Syndicate'. At one time President F. D. Roosevelt proposed that the surplus Chinese population should be settled in the region and Amazonia was also proposed as a place where the Jewish people could set up a new State of Israel.

In 1967 the Brazilians were alarmed by the proposal of Herman Kahn and Robert Panero of the Hudson Institute in New York that a series of dams and gigantic lakes be built in Brazil, Colombia, Peru, and Bolivia to flood Amazonia and make it into a massive fisheries-production zone. This proposal, made without thought for the Amazon biodiversity, its native peoples, or its mineral deposits, naturally fuelled Brazilian concern about foreign interference with its territory. It was probably this Hudson Institute project that precipitated the decision of the military government of Brazil to build the Transamazon Highway and occupy the Amazon definitively. The development of the region and of modern industrial Brazil by the military also brought to Brazil an enormous foreign debt. With a $50 billion service charge owed to international banks just to maintain its debt, without paying off the capital, the 'paranoia' continues because it is the debt that is now ruining Brazil's ability to develop or to conserve Amazonia.

National concern about Amazonia continued to be fuelled by statements that Brazil must accept limited sovereignty status over the Amazon that have been made in recent years by US

Senator Robert Kasten and by the French President François Mitterand. Thus today when pressure is put on Brazil by many foreign nations to recognize its limited sovereignty over the territory of the Yanomami Indians or to preserve the whole Amazon region as a biological reserve it is hardly surprising that she is indignant. It is logical that Brazil is indignant when the USA or Britain tells her to stop cutting forests because the carbon dioxide released contributes to the greenhouse effect, when we are reluctant to economize on the fossil fuels that we burn. It is against this background of national paranoia, which has considerable historical justification, and an overburden of debt that has been encouraged by the developed world, that one must consider the problems of the Amazon region.

ECOLOGICAL BACKGROUND

In order to understand the dilemma of the conflict between conservation and utilization of the Amazon rainforest, it is also necessary to understand some of the basic ecological constraints and considerations that must be taken into account. There are many aspects that could be discussed, but the three most important are probably species diversity, soil poverty, and interactions between the species.

Species diversity

Unlike the forests of the temperate regions, where a few or even a single tree species dominates the forest, most tropical rainforests are composed of a great variety of species mixed together in low population densities. The highest recorded species diversity in rainforest was found at Yanomono near to Iquitos, Peru by Alwyn Gentry of the Missouri Botanical Garden (Gentry 1988) who found 300 species of trees and lianas of 10 cm diameter or more in a single hectare. Some samples of Amazonian forest diversity are given in Table 7.1. To the developer this diversity is a problem because individual species

Table 7.1 Tree species diversity for some tropical rainforests.

Locality	Number of species	Minimum diameter (cm)	References
Yanomono, Peru	300	10	Gentry (1988)
Mishana, Peru	295	10	Gentry (1988)
Yasuni, Ecuador, T.f.	228	10	Balslev *et al.* (1987)
Johore, Malaysia	227	10	Prance (1990*b*)
Mulu, Sarawak	223	10	Prance (1990*b*)
Cocha Cashu, Peru	189	10	Prance (1990*b*)
Manaus, Brazil	179	15	Prance *et al.* (1976)
Tambopata, Peru	168	10	Gentry (1988)
Xingu River, Brazil (3 separate hectares)	162, 133, 118	10	Campbell *et al.* (1986)
Breves, Brazil	157	10	Pires (1966)
Yasuni, Ecuador, floodplain	149	10	Balslev *et al.* (1987)
Oveng, Gabon	131	10	Prance (1990*b*)
Alto Ivon, Bolivia	94	10	Boom (1985)
Belém, Brazil	87	10	Black *et al.* (1950)

occur with such low population density that they are hard to exploit. It appears to be much easier to cut the original forest and replace it with a monoculture of pasture grass or commercial timber. However, such monocultures rarely thrive in the tropics because of their susceptibility to disease and predator attack. To the conservationists the diversity is what makes the forest interesting and its preservation vital. Because of low population densities and the localized distribution of many species, it is necessary to preserve large areas to have an effective conservation policy. The conflict between use and conservation begins with the problems of how to cope with a species-diverse forest.

Soil

It is often presumed that because of its luxuriant growth, rainforest must lie over a rich soil. Quite the contrary is the case for most of the Amazon rainforest, which covers areas of extremely poor soil with low nutrient content (Table 7.2). Some of the Amazon forest is over sandy soil that contains virtually no nutrients. The available nutrients are in the vegetation rather than in the soil, and they are rapidly recycled as leaves and branches drop to the forest floor. It is common to see roots of living trees growing upwards in search of nutrients into dead trunks and tree-stumps. The tree roots are superficial, and rotting leaves are often linked to them directly by mycorrhizal fungi, which are some of the most important organisms in the forest because of the role they play in the nutrient cycle. The poorer sandy soils have a greater proportion of the trees with a mycorrhizal connection than the richer clay soils (St John 1985).

If the forest is cut and burned, then most of the nutrients are washed away into the streams by the heavy rainfall. Streams in undisturbed rainforest are of pure, almost distilled water, but streams in deforested areas are full of the escaping nutrients

Table 7.2 Basic properties of Amazonian soils comparing those of floodplain, forest on terra firme, white sand areas and cattle pasture. Analysis made at INPA, Manaus by F. Magnani, 18 September 1982.

Sample	Habitat	pH	P(ppm)	K(ppm)	Ca	Mg	Al
401	Ariau—Várzea floodplain	5.3	70	45	7.0	2.4	0.3
402	Ariau—Várzea floodplain	5.3	85	62	8.5	2.3	0.3
403	T. Loureiro cattle pasture	4.0	2	12	0.3	0.1	2.0
404	T. Loureiro cattle pasture	4.0	3	20	0.5	0.2	2.0
405	Inpa white sand campina	4.7	2	8	0.4	0.2	0.7
407	Inpa terra firme forest (KM45)	3.5	1	8	0.3	0.1	2.8
408	Inpa terra firme forest (KM45)	3.5	1	12	0.3	0.1	3.0
409	Curuá-Una floodplain forest	6.6	65	46	4.5	0.8	0.0

Agricultural needs for nutrients analysed:

  K low—10 —middle— 30 high
  P low—46.8—middle—117 high
  Ca low—2.0 —middle—5.0 high
  Mg low—0.5 —middle—1.0 high
  Al: above 0.5 is often toxic to plants

that are vital to sustain the forest or even the crop. The clay latosols and sandy soils of Amazonia do not have the colloidal properties to retain nutrients. When the latosol is converted to cattle pasture its drainage pattern is easily changed by the compaction caused by cattle hooves. The prudent management of the majority of Amazonia which is covered by poor soils should neither break down the delicate balance of the nutrient cycle nor compact the soil. The ecology of the region points towards species-diverse forest as the best land cover, and plans for its use and development should be geared towards maintenance of the forest cover.

However, not all Amazon soils are poor, for example see Jordan and Herrera (1981) and Lathwell and Grove (1986). It is just unfortunate that most major development projects have taken place in regions with poor soil. The need to be selective is touched on in the section about the richer alluvial soils of the floodplain (p. 179). Cochrane (1984) estimated that only six per cent of the Amazon terra firme has well-drained and fertile soils, so there is not much room for choice.

Species interactions

To manage or to conserve the forest requires a knowledge of the way in which one species affects another. Recent biological research has given us numerous examples of animal–plant interactions in rainforest (for some examples see Kubitzki 1983, 1985; Pannell, 1989; Prance 1985). A single example from my own research will illustrate the way in which unexpected organisms are linked together. The Brazil nut (*Bertholletia excelsa* H. and B.) is an important economic product of Amazonia. The flowers of the Brazil-nut tree are pollinated by various species of large bees that are capable of lifting the tightly closed cap that covers the pollen-bearing stamens. The commonest pollinator is the Euglossine bee *Eulaema meriana* which carries pollen from one tree to another and effects the cross-pollination that is essential for fruit set. The males of this same bee depend on various large-flowered species of orchids from which they gather scents to attract female bees for mating to take place (see Dodson 1966, 1967; Dressler 1968a, 1968b; Nelson et al. 1986; Prance 1976, 1985; Mori and Prance 1990). Thus, the Brazil-nut pollinator requires an orchid species to survive. Brazil nuts grow in arrangements like the segments of an orange, in large round woody fruit capsules which fall to the forest floor when they are ripe, 14 months after pollination. The only animal capable of chewing open the hard outer fruit cover to remove the nuts is a rodent, the agouti (*Dasyprocta*). These animals take the nuts away from the parent tree and bury them in hoards, some of which they forget. The dispersor of the seeds of the Brazil-nut tree is therefore the agouti. Thus we see that the economically important Brazil nut is interdependent at least with bees, orchids, and agoutis. Each of those organisms is in turn linked to other species within the forest. The whole species-diverse forest is linked together in a vast and complicated web of plant–animal interactions. To manage the Brazil nut it is necessary to understand and conserve the species with which it interacts, and this is true of any species within the forest. Early attempts to grow Brazil nuts in plantations failed because the pollination system was not understood. Use of the rainforest species will often depend on the conservation of many other species upon which they depend, whether it be a bee to pollinate the flowers or a mycorrhizal fungus to assist the absorption of nutrients from the leaf litter.

THE SHORT-TERM UNSUSTAINABLE APPROACH

In Amazonia the greatest single cause of deforestation has been the creation of cattle pasture which, as we have already seen, is unlikely to work because of the poor soils that compact easily. Cattle pasture was only made viable by huge government tax-incentives and by land-price speculation. There were even some ranches that never sold cattle to the market; the tax break was sufficient. Most Amazonian cattle pasture is totally unproductive, sacrificing the high productivity of the rainforest to raise less than one cow per hectare. Time has also shown that pasture is not sustainable over a long period since poisonous weeds soon take over. The creation of cattle pasture owed much to the political motives of a military government that needed to open up the Amazon region to stake claim to the territory. No notice was taken of the strong scientific evidence against the wisdom of this use of the land. 1987 was the year of greatest deforestation, when 50 000 square kilometres were lost as landowners sensed the implementation of control and an end to tax incentives (see Setzer *et al.* 1988). There was a rush to take ownership of the land by felling the trees. Fortunately, however, I do not need to develop this theme any further because current federal government policy in Brazil has completely reversed the trend. Since 1988 farmers who cut down the rainforest illegally are being fined. The amount of Brazilian Amazon forest cut and burned in 1989 was 21 000 square kilometres. The recently released figures for 1990 (*Veja*, 9 Jan. 1991) estimate that only 10 000 square kilometres of forest were destroyed and the number of individual centres of fires was reduced by 25 per cent. This is encouraging news indeed for which the Brazilian government should be congratulated. The tragedy is that so much forest was destroyed for a non-sustainable use and from purely political motives.

The most serious threat to Amazonia today is the invasion of the region by hoards of goldminers who have occupied the headwater regions of most of the major Amazon tributaries, especially in Roraima, the home of the Yanomami Indians. In that state 45 000 miners infiltrated the Indian territory and brought Western diseases that have already greatly reduced their population. The greatest cause for concern is that the miners use mercury to separate the gold. The mercury is burned off and released to the environment and has now become the most serious contaminant of the Amazon region. Sir Richard Southwood in his inaugural Linacre Lecture (this volume, chapter 1) emphasized the problem of lead and other heavy metal poisoning and its cost to Western society. Amazonia is only just beginning to realize the cost of this uncontrolled release of mercury into the ecosystem. It is again the rush for a short-term profit regardless of the environmental consequences that has led to this problem. The Brazilian government has taken seriously the eviction of miners from Yanomami Territory, but epidemics of gold fever can break out anywhere and do much harm before they are controlled.

Some experiments have been made in Amazonia with the creation of large forestry monocultures. These have also largely been failures. The two best-known examples are Fordlândia, the rubber plantation created by Henry Ford, and the Jarí forestry project founded by multimillionaire Daniel Ludwig. In 1926 the Ford Motor Company decided to plant a huge rubber-tree plantation in the Rio Tapajós region of lower Amazonia at Fordlândia and Belterra, to create a major source of rubber in its native Amazon region. The project was a complete failure because of the number of diseases that attack rubber in its native territory, especially the leaf-rust fungus *Microcyclus ulei*. Fortunately for the world, the diseases were not taken to Asia and Africa when rubber was planted there and so the plantations thrive. Wild rubber trees are spread sparsely throughout the Amazon forest, and are not badly attacked by fungal diseases because they do not spread easily from tree to tree in species-diverse rainforest—transplant rubber to a plantation in Amazonia and diseases jump quickly from tree to tree. Consequently Fordlândia, and many other attempts to cultivate rubber in Brazil, failed. The Jarí project was the dream of Daniel Ludwig who forecast in the 1960s a future world shortage of paper pulp. He decided that the fast-growing tropical relation of teak, *Gmelina arborea*, would be the solution. After exploring various possible places such as Nicaragua to locate his vast plantation, Ludwig chose the Jarí river which forms the borders of the states of Pará and Amapá in Brazil. The project was started without any preparatory research in 1967, and large areas of the original forest were felled to be replaced by young gmelina trees. The clearing was carried out by bulldozers which removed all the debris'and top soil. It was soon noted that the trees were not growing as expected because of the poor soil that remained. Consequently gangs of axe-men were brought in to replace the bulldozers and the topsoil was left intact. The gmelina grew better amongst the debris of the former forest rather than on bare, eroded soil. However, it only grew in those areas of the property which had clay soils. The Jarí plantation area borders the clay alluvial soils of recent Tertiary and Quaternary origin and the ancient sandy soils of the Guiana Shield. The basic ecological research to determine the soil preference of gmelina had not been performed, and in plantations on sandy soil gmelina had to be replaced by slower-growing pine trees (*Pinus caribea*). In spite of an investment of over a billion dollars, Jarí has not proved economically viable. Mr Ludwig sold out in 1982 to a consortium of 27 Brazilian companies backed by their government. His loss was 600 million dollars. The plantation continues and is producing pulp, but only breaks even because of the deposit of kaolin on the property which is being mined profitably (for further details about Jarí see Fearnside 1988; Fearnside and Rankin 1980, 1982, 1985).

Many other examples could be given of the deforestation of the Amazon by projects which are either uneconomic or non-sustainable over an extended time period. Until recently this lack of sustainability was true of almost all the efforts to develop the region that involved deforestation. However, the purpose of this lecture is not to dwell on the disasters, but to examine the alternatives to deforestation, to discover Southwood's 'Best Practicable Environmental Option'.

TOWARDS SUSTAINABILITY

The factors that affect the development of the Amazon region include history and sovereignty of the territory, the fragile ecology of the region, and the need to sustain the population in a viable manner. It is obviously a given factor in the equation that the entire Amazon could not be set aside as a vast biological reserve to preserve evolutionary and cultural history in a pristine environment. However, our accumulated knowledge both from failures and from successes shows that there could be a middle road that combines presentation of much of the forest together with a reasonable livelihood for an Amazonian population of a reasonable size. I will present some of these options, which are based on the results of our research in ecology and economic botany. Since we have seen that the neglect of ecological principles leads to such failures as Jarí and much of the cattle pasture, it is certain that a viable land-use system can only be achieved if it is one that takes into consideration the basic ecological facts.

The maintenance of the forest cover is necessary not only for sustainable systems of utilization, but also to preserve regional and world climate stability. Salati *et al.* (1986), and Salati and Vose (1984) have shown that almost 50 per cent of the rainfall in the Amazon forest is the result of evapotranspiration from the trees, and another 25 per cent is evaporated off the canopy. That is, only a quarter runs off to the rivers. The oreographic rainfall from the east coast falls in eastern Amazonia. Much of it is transpired through the trees or also evaporated and this water vapour is carried further west by the prevailing winds. If forest cover is removed and replaced by vegetation types with less leaf surface, less water will be recycled and more will run off into the rivers. Deforestation in Amazonia will cause a severe disruption of the rainfall patterns regardless of the additional contribution it would make to atmospheric carbon dioxide and the greenhouse effect (see also Shukla *et al.* 1990). Sir Richard Southwood has already pointed out in Chapter 1 that destruction of the rainforests contributes to the increase in carbon dioxide and has explained the greenhouse model (see also Myers 1990). There are, therefore, many compelling reasons why the Amazon region should remain covered by forests. However, Amazonia covers 60 per cent of the territory of Brazil and a considerable proportion of the territory of seven other countries who cannot be expected to leave the forest untouched, yielding nothing to their economy except ecological stability.

170 The dilemma of the Amazon rainforests

Realization of the inefficiency of much of the deforestation, further research, and a greater public concern is leading the Amazonian countries to search for alternatives to deforestation.

ALTERNATIVES TO DEFORESTATION

The forest Indians' system

Recent studies in quantitative ethnobotany (Boom 1985, 1987; Balée 1986, 1987; Prance et al. 1987; Milliken 1991) have shown the extent to which the Amazonian Indians use the range of tree species in the rainforest. Boom (1985, 1987) found that the Chácobo Indians of Bolivia have uses for 82 per cent of the species of trees on a sample hectare (75 species used out of a total of 91), representing 95 per cent of the individual trees (619 out of 649). Table 7.3 gives a summary of these uses. A similar study by William Balée (1986) showed that the Ka'apor Indians of eastern Amazonian Brazil have uses for every single species of tree and vine of 10 cm diameter or greater on the hectare studied (135 species). The studies in quantitative ethnobotany which have all been made during the last eight years show varying levels of use of rainforest species from 50 per cent (the

Table 7.3 Uses of plants by the Chácobo Indians of Alto Ivon, Bolivia on a hectare of forest with 91 species and 649 individual trees. (Data from Boom (1987))

Use	No. of species	No of individuals	Percent of total species
Commercial (rubber latex)	1	5	1.1
Fuel	14	163	15
Medicinal	23	271	22
Construction and crafts	23	225	22
Food	33	264	36

Panaré of Venezuela, Boom 1990) to 100 per cent in the case of the Ka'apor. These studies show only the use of trees; in addition the Indians use many of the herbs, shrubs, epiphytes, ferns and even mosses, lichens, and mushrooms. The plants of the forest provide their food, fuel, construction materials, basketry fibres, medicines, clothes, dyes, and many other products. It is not in the interest of people who depend on the rainforest to such an extent to cut it all down. The tribes mentioned above are all agriculturists as well, but they cut and burn only small areas because so much of their sustenance depends upon the forest. One way to preserve the rainforest is to follow the example of the Indians through extraction of products without destruction. Areas that are Indian reservations are notable for their forest cover and most species have been preserved. The only noticeable problem is the over-hunting of certain game animals.

Extraction forests

Timber extraction from the rainforests of the world has generally been carried out through the use of unsustainable methods. An excess amount of timber is removed and great damage is caused to the rest of the forest. The International Tropical Timber Organization (ITTO) has estimated that less than an eighth of one per cent of all timber harvested from the rainforest is extracted by truly sustainable methods. The route to sustainable use of the forest must be through reducing the amount of timber extracted and increasing the number of non-timber products that are extracted from the forest.

In Amazonia many of the settlers have made their living through the extraction of non-timber products such as rubber latex, Brazil nuts, balata, and the latex used for chewing gum, resins, and oils. Worldwide attention focused upon the rubber tappers of Brazil when one of their leaders Francisco (Chico) Mendes, President of the Rubber Tappers Union of the State of Acre was assassinated in December 1988. Mendes led the resistance by rubber tappers to the deforestation of their forests by wealthy cattle ranchers. His death drew so much international attention that, unlike previous similar cases of murder, the culprits were tried and jailed and the cause of the rubber tappers was given new impetus. As a result Acre, and later some of the other Amazonian states of Brazil, have set up 'extractivist reserves' where people are allowed to live and use the forest but not to cut it down. This concept of extractivist reserves is developing well (see Allegretti 1990, and Fig. 7.1), but so far it has been based largely on two products: rubber latex which is tapped in the dry season and Brazil nuts that are gathered in the rainy season. The price of rubber in Brazil is maintained above the world market price by subsidies and Brazil nuts are starting to be grown successfully in plantations. Extractivist forests will only be viable in the long term if they are based on a diversity of products, like the Amazonian Indian system which is based on a diversity of species. The challenges for the future are to discover non-timber forest products which can be harvested sustainably from the forest, and to create a market for them. The Director of Cultural Survival, Dr Jason Clay, has begun such a search for products and has interested various companies in their marketing. The candy 'Rainforest Crunch' is already popular in the USA and is now to be marketed in the UK by the Body Shop, a company interested in developing rainforest cosmetics. Economic botanists must now help in the search for new rainforest products, and in the creation of a market for them. To be sustainable these are most likely to be products made from latex and resin that can be tapped from the trees, or products from the fruit and seeds of trees. For example, the timber tree andiroba (*Carapa guianensis*) also produces a large fruit from which the Indians extract a medicinal oil. Could andiroba oil be used as a base for cosmetics? The trees produce an annual crop of seeds and need not be felled to extract this product. The copaiba tree (*Copaifera*) produces an oleoresin that has long been tapped from its trunk and already has a limited market. Can we develop new uses for this product? There are numerous possibilities, and extractivist forest reserves will only be sustainable themselves if they become truly viable in economic terms and are able to provide more than a subsistence livelihood.

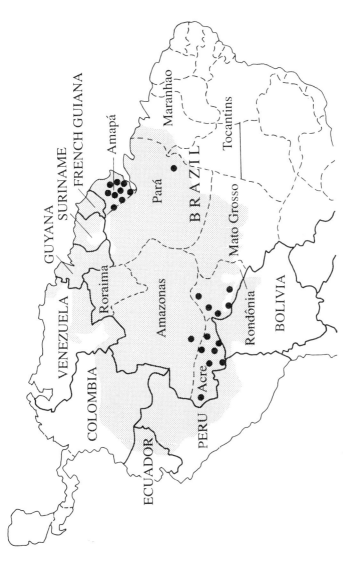

Fig. 7.1 Location of extractive reserves (shown by black dots) that are proposed or implemented in Amazonian Brazil. These are mostly located in the states of Acre in the west and Amapá in the east. The tinted area shows the limits of the Amazon rainforest.

Indigenous agroforestry systems

Agroforestry which combines growing mixed tree and herbaceous crops is becoming an option for use of some of the fragile tropical soils, since it avoids excess erosion. Although agroforestry is a comparatively recent term, it has long been practised by indigenous peoples, especially in rainforest areas. A detailed study has been made of the agroforestry system of the Bora Indians of Peru (Padoch *et al.* 1985; Denevan and Padoch 1987). An interdisciplinary team studied a series of former fields of different ages from three to nineteen years since felling. The results (see Figs 7.2–7.4) show that the fields are initially planted with a mixture of herbaceous and woody crops arranged in small patches. Crops are interspersed rather than planted in monocultures, and areas that initially produce manioc or cassava gradually become small orchards of fruit trees such as umari (*Poraqueiba sericea* Tul.). At any moment in time the Bora have a series of fields and forests of different ages from which they can harvest different products. As natural regeneration takes place and as the fields return to forest they still yield food, fibre, medicines, and other products used in the daily life of the tribe. For example, a 19-year-old forest contained 22 useful species as well as a small orchard of the edible fruit macambo (*Theobroma bicolor*).

A similar system has been created by the residents of the small Peruvian town of Tamshiyacu. They use the same crops as the Bora Indians, and go through 35-year cycles of land use between each clearing. The system yields a good income for the residents of the town because of the proximity of the market town of Iquitos. It produces a great variety of crops for their daily needs and thus avoids the necessity of buying many products. For example, the oil from the seeds of *Couepia dolichopoda* provides a cooking oil which can also be burned in lamps. One of their most marketable fruits is the umari fruit, and one finds small orchards scattered within their agroforestry areas. When the system is to be felled the umari trees are used to make another highly marketable product, charcoal. Brazil-nut

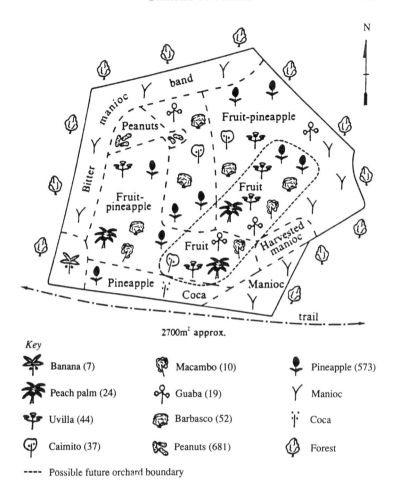

Fig. 7.2 A three-year-old Bora Indian field showing their species-diverse agroforestry. (Adapted, with permission, from Denevan and Padoch 1987, *Advances in Economic Botany*, The New York Botanical Garden.)

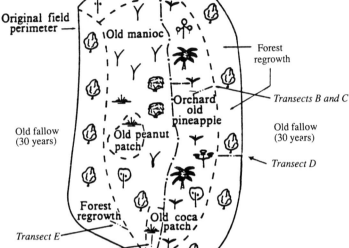

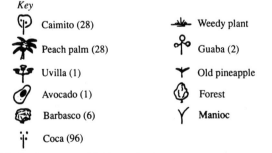

Fig. 7.3 A five-year-old Bora Indian transitional orchard fallow. (Adapted, with permission, from Denevan and Padoch 1987, *Advances in Economic Botany*, The New York Botanical Garden.)

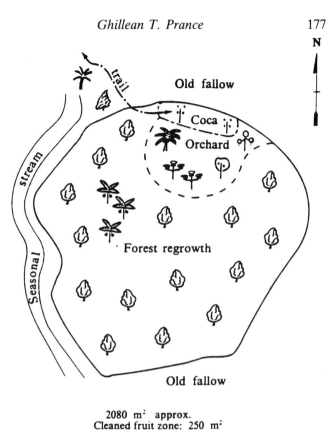

Fig. 7.4 Nine-year-old orchard fallow. (Adapted, with permission, from Denevan and Padoch 1987, *Advances in Economic Botany*, The New York Botanical Garden.)

trees are also planted at Tamshiyacu and when the forest is recut, some trees are left standing as shade and to continue productivity into the next cycle, whilst others furnish an excellent timber for house-building.

Several other indigenous and settler systems of agroforestry in Amazonia have now been studied. The most detailed study is of the Kayapó Indians of Pará who use both rainforest and savanna (see Posey 1982, 1983, 1984). Anderson and Jardim (1989) studied the exploitation system of settlers in the Amazon delta region, and showed that they make an excellent living from managing the açaí palm (*Euterpe oleracea* Mart.) to provide heart-of-palm and fruit, tapping rubber trees and exploiting various other products. The ecological profiles of areas of exploited forest are remarkably similar to those of the natural forest of that region.

One of the strengths of the Bora Indian and Tamshiyacu agriculture systems is the use of a large number of varieties of each of their main crops. There are several named varieties of umari and a larger number of cassava. The use of these varieties maintains adequate genetic diversity in their crops which makes them less prone to disease and predators and more flexible for improvement. All the groups of Indians I have studied use many varieties of manioc. Chernela (1986) found 137 cultivars amongst the Tukano Indians of the upper Rio Negro, Boster (1984) described 50 cultivars in use by Jivaro Indians of Peru, and Carneiro (1983) also listed 50 cultivars for the Kuikuru Indians of Brazil. The exploitation of such genetic diversity is normal in indigenous agriculture. The tendency of modern agriculture is for uniformity of crops both genetically and phenotypically. The systems that maintain genetic diversity have a much greater long-term future in the tropics.

The agricultural system of the Bora and Kayapó Indians and of various settlers who have adopted similar methods are based on species diversity, genetic diversity, and on soil conservation (Hecht (1989) describes soil management by the Kayapó). They are much more likely to have long-term success because they are closer to the natural ecology of the region. The examples which we already have are enough to indicate that where the forest must be replaced to sustain the population it will be much more viable to use agroforestry rather than monoculture. This brings the additional benefit of a reasonably high biomass and consequently the retention of carbon in the vegetation and the soil rather than its release to the atmosphere.

Floodplain agriculture

In addition to their system of agroforestry on the upland terra firme forest, the residents of Tamshiyacu grow rice on the muddy banks and exposed fields beside the river. The river levels of Amazonia fluctuate annually between the rainy and the dry season. During the dry season muddy areas are exposed that are ideal for cultivation of fast-growing crops such as rice or maize. The sale of rice is an important part of the economy of Tamshiyacu and other riverside communities.

It has frequently been proposed that the floodplain of the silty white-water rivers be used for agriculture. For example, pioneer Amazonian ecologist Felisburto Camargo (1951) was a great proponent of the use of the *várzea*, as it is called (see also Goodland and Irwin 1975). Indeed the passenger of a cruise ship passing along the lower Amazon will note that much of the riverside vegetation has been replaced by agriculture and cattle and water buffalo pasture. A crop that has helped the riverside dweller is jute, introduced from Asia. It grows well on the floodplain and jute factories have been built in several Amazon cities such as Manaus and Parintins.

Table 7.2 (samples 401, 402, and 409) shows the comparative nutrient richness of floodplain soil as well as its lack of toxic aluminium. These data should be enough to indicate that the floodplain is more appropriate for agriculture than the terra firme. The soil of the floodplain is replenished annually by a rich deposit of alluvial matter.

However, a word of caution is needed. It would be most unfortunate if the entire area of várzea forest were to be replaced by agriculture, because there are species that are endemic to várzea which would become extinct if all of the original forest was to be felled. In addition there is a close relationship between the fish and the várzea forest. Many Amazonian fish are vegetarian and swim through the forest in the flood season to eat the fruit of many species of trees. The fish develop fatty tissue during the bounty of the floods and then return to the riverbeds during the dry season and live primarily off their reserves of fat (see Goulding 1980, 1989; Smith 1981). The Indians bait their fish-hooks with the fruits of the jauari palm (*Astrocaryum jauari*), the *Alchornea* shrub, and rubber, which are all floodplain plants that drop their seeds into the water during the flood season. I have often watched fruit falling from a *Licania* tree being ingested immediately by fish as they plop into the water. To lose floodplain forests would be to destroy one of the most valuable resources of the region, the fisheries, which like the timber are already being seriously overexploited. The use of the várzea forests for agriculture can only be in moderation. It is possible to create floodplain agroforestry systems that incorporate useful species such as rubber, jauari palm, and andiroba that are also favourites of the fish. As with many Amazon problems, a balance between use and conservation is needed in the floodplain. The stability of floodplain areas is also an important consideration. They are already constantly changing through the process of erosion and building up of mud banks, and deforestation increases erosion and loss of the areas suitable for cultivation.

Oligarchic forests

This is a recently coined term for forests that are dominated by a single species of tree (Peters *et al.* 1989). Even within the species-diverse Amazon forest, there are certain areas where the stress caused by some environmental factor limits the number of species that can grow there. This is especially true of swamps. Oligarchic forests are naturally occurring single-species stands that resemble monocultures. Fortunately the species that dominate oligarchic forests are often some of the most useful to humans.

Large areas of Amazonian swamp are dominated by the aguaje or buriti palm (*Mauritia flexuosa* Mart.). It occurs in waterlogged areas throughout Amazonia and has been put to many uses. In the upper Amazon the fruits are a popular food and ice-cream flavour and are much sold in the markets of Iquitos, Peru. In the lower Amazon at Maranhão, Brazil, much of the local craft, such as hammocks, baskets, and tablecloths is made from the fibre extracted from the epidermis of the young leaves of mauritia, and the fruit is made into a sticky paste for use as a dessert. The fruit is rich in vitamin A and is an important dietary item. Since this species occurs in large natural stands and it is largely the fruits that are exploited, it would seem an ideal crop for further development. However, wherever humans exploit a resource they tend to abuse it rather than plan for the future. The Peruvians usually fell the mauritia trees to collect the fruit. There are now many stands of the palm where only male trees remain because all the fruit-bearing female trees have been cut down. This is quite unnecessary because, even though trees are up to 30 metres tall, the fruit could be collected either by climbers or with the use of clippers on poles.

Another oligarchic forest dominated by camu-camu (*Myrciaria dubia* (HBK) McVaugh) occurs by lakesides and riversides of the upper Amazon. The camu-camu shrub frequents a narrow belt from the water's edge to the beginning of the várzea forest in an area that is under water for ten months of the year. The camu-camu is one of the few species that can withstand this prolonged flooding. The shrubs flower and set fruit during the dry period and by the time the water level rises the cherry-like fruits are ripe. They contain up to 30 times more vitamin C than does citrus. The fruit is easy to harvest from canoes as river-levels and lake-levels rise. In recent years camu-camu has become quite a popular fruit in Iquitos and it has considerable potential if a wider market could be developed. The ecology and productivity of the species have been studied by Peters and Vasquez (1987), and Peters and Hammond (1990) who found that since it forms dense stands it is an ideal candidate for sustainable harvesting.

A third example of an oligarchic forest occurs in the transition region between eastern Amazonian rainforest and the central Brazilian savanna or Cerrado. This region is dominated by the useful babassu palm (*Orbignya phalerata* Mart.; see Balick 1986). This species is fire resistant and so survives the fires that spread from the savanna boundaries. The quantity of babassu forest has also been vastly increased by deforestation of the transition forests. Babassu has many uses, especially the kernel of the fruit which has been used by local peasants as a source of cooking and lighting oil. The outer mesocarp of the fruit is rich in carbohydrate and can be used as flour, and the endocarp that surrounds the kernels is hard and woody and is used to make a high-grade, sulphur-free charcoal. The oil is used industrially for cooking, soaps, and other uses. Another excellent change towards sustainable use of resources has been made by the CERMA beer factory of São Luís in Maranhão. Instead of continuing to use firewood from the coastal mangrove forest and ruin the fisheries, the factory has converted its boilers to use babassu-palm fruits. A mixture of 90 per cent cracked-open shells and ten per cent whole fruit with the kernels full of oil to increase burning temperatures makes a fine boiler fuel. The babassu produces annual crops of fruit and so a sustainable source of fuel has been found to substitute for the wood of the mangrove trees.

The further use of these single-species-dominated forests could take the pressure off the use of the species-diverse forests. The oligarchic forests of Amazonia contain an abundance of products because many of them consist of multiple-use palm species.

The use of extraction from the forests, agroforestry, floodplain agriculture, and oligarchic forests are all part of the pathway from destruction to the sustainable use of the Amazon forests. Many other methods and examples will be created and timber extraction at a sustainable level is certainly part of the future. In addition the Amazon region is rich in mineral resources, but these must be exploited with greater ecological care. Already some of the large mining companies working in the region have extensive and successful programmes of rehabilitation after mining. Fine examples are at the bauxite mine of Mineração Rio Norte on the Trombetas river and the iron mine of the Carajás project (see Prance 1989).

CONSERVATION

The change from unsustainable to sustainable land use in Amazonia will reduce the amount of deforestation necessary, but even sustainable systems will alter the species composition, reduce the availability of soil nutrients, and cause ecological change. For example, the hunting of animals depletes the stock in extractivist reserves. It is vital that an adequate system of biological reserves be maintained to preserve the diversity of the species of Amazonia and to make the land-use systems described above a viable proposition. On paper there are already a large number of forest reserves, parks, and indigenous reserves within Amazonia (Fig. 7.5). However, few of the reserves are in any way protected and settlement has already occurred within some. Recently the number of reserves has been increasing and the return of land title to the Indians of an area twice the size of the United Kingdom by President Barco of Colombia is encouraging. Even if the current nominal reserves were to be upheld, they would not conserve nearly enough land to preserve the Amazonian ecosystem intact. If the climate of the region is to be maintained, climatologists such as Eneas Salati, Director of the National Amazon Research Institute in Manaus, indicate that about 80 per cent of the forest must remain. Some of this can take the form of extraction forests, plantations, and agroforestry systems; but a considerable area must be set aside as biological reserves. To maintain the species of Amazonia in viable populations it will be necessary to set aside at least 20 per cent of the area as biological reserves. These must be carefully selected to include the many different types of vegetation and the clusters of endemism of plants and animals.

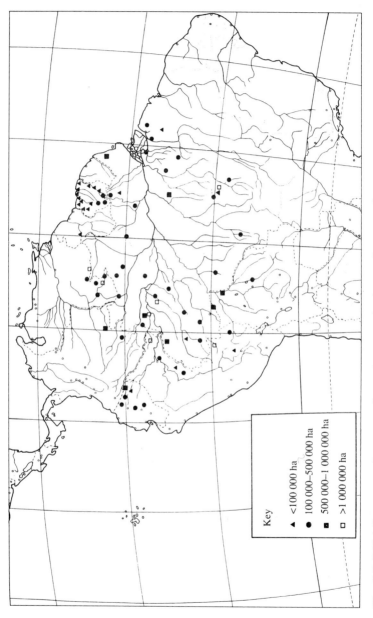

Fig. 7.5 Officially protected areas, parks, and reserves in Amazonian countries.

A workshop was held in January 1990 in Manaus to identify areas of biological priority for conservation (Prance 1990*a*); 100 biologists and other scientists gathered together to pool their data to produce a consensus map of the areas that are most important for the conservation of the biodiversity of Amazonia. Five levels of priority were given and the areas considered worth conserving covered about 60 per cent of the region, with a much smaller area given the highest priority. The resulting map (Fig. 7.6) gives a scientific rationale for the creation of conservation areas within Amazonia. The data from Workshop 90 have been given to the conservation organizations of each Amazonian country and are being considered by some. Already two reserves have been created in the state of Amazonas, Brazil, in priority areas suggested at the workshop.

Conservation of Amazonia has a long way to go, but at least Workshop 90 has provided data for planners to choose appropriate areas for reserves and for politicians to make them a reality. Reserves must be placed in areas that conserve species and ecosystems rather than in areas that are selected because the land is unsuitable for other uses, as has been the tendency so far. Reserves are also necessary to protect the wild relatives of Amazonian crop plants such as rubber, cacau, and mahogany because the future of these valuable species may depend on the genetic diversity of their closest relatives.

CONCLUSION

The Amazon region is at a crossroads today and there are many forces pulling in different directions. On the one hand, developers and rich landowners pay little heed to conservation and are interested in short-term profit, and on the other hand ecologists, conservationists, indigenous peoples, and some enlightened planners and politicians, such as José Lutzenberger, Environmental Secretary of Brazil, and ex-President Barco of Colombia, understand the scientific facts about the fragile ecosystem and are the visionaries upon whom the future of Amazonia depends.

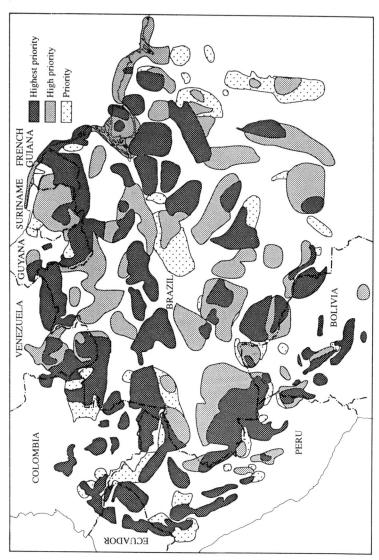

Fig. 7.6 Conservation priorities for the preservation of biological diversity from the results of Workshop 90 in Manaus, Brazil.

It is not too late to change course because only ten per cent of the region has been deforested, but the efforts of President Collor of Brazil and his Environment Secretary Lutzenberger are being strongly resisted. Several governors of Amazonian states were elected in the November 1990 elections on clearly anti-ecological platforms. For example, Gilberto Mestrino of Amazonas declared in his election campaign that he would not be the governor of alligators and trees, but of the people: if it was necessary to destroy the trees and the alligators, he would do so for the people. If he were to fulfil his promise the destruction would probably also destroy the livelihood of the people. At present there is a battle going on in Amazonia between such people as Governor Mestrino and President Collor. A future for the region can only be assured if a balance between conservation and utilization is achieved in which the region becomes neither an untouchable biological reserve nor a devastated Sahara desert. Ecologists, climatologists, soil scientists, economic botanists, foresters, agroforesty experts, and conservationists are beginning to accumulate enough data to develop the region sustainably. It is to be hoped that enough political momentum can be generated to allow them to implement their ideas. The colonialist intrusions of the past must be laid aside, and the justifiable paranoia of the Amazon countries ended so that we can work together to preserve both the biodiversity and the sovereignty of the Amazon countries, and thereby give them a sustainable future.

ACKNOWLEDGEMENTS

I am particularly grateful to David Guieros Vieira for furnishing information for the historical section and to the Instituto Nacional de Pesquisas da Amazônia and the Museu Paraense Emílio Goeldi for much collaboration during fieldwork in Amazonia.

REFERENCES

Agassiz, L. and Agassiz, E. (1969). *A Journey in Brazil*. Praegor, New York.

Allegretti, M. H. (1990). Extractive reserves: an alternative for reconciling development and environmental conservation in Amazonia. In: *Alternatives to deforestation* (ed, A.B. Anderson), pp. 252–64. Columbia University Press, New York.

Anderson, A. B. and Jardim, M. A. G. (1989) Costs and benefits of floodplain forest management by rural inhabitants in the Amazon estuary: a case study of açaí palm production. In: *Fragile lands of Latin America: the search for sustainable uses* (ed. J. Browder), pp. 14–26 Westview Press, Boulder, Colorado.

Balée, W. (1986). Análise preliminar de inventário florestal e a etnobotânica Ka'apor (Maranhão). *Boletim do Museo Paraense Emílio Goeldi Sér. Botânica*, 2(2), 141–67.

Balée, W. (1987). A etnobotânica quantitativa dos índios Tembé (Rio Gurupi, Pará). *Boletim do Museo Paraense Emílio Goeldi Sér. Botanica*, 3(1), 29–50.

Balick, M. J. (1986). Systematics and economic botany of the *Oenocarpus-Jessenia* (Palmae) complex. Advances in Economic Botany 3.

Balslev, H., Luteyn, J., Ollgaard, B., and Holm-Nielsen, L. B. (1987). Composition and structure of adjacent unflooded and floodplain forest in Amazonian Ecuador. *Opera Botanica*, 92, 37–57.

Black, G. A., Dobzhansky, T. and Pavan, C. (1950). Some attempts to estimate species diversity and population density of trees in Amazonian forests. Botanical Gazette, 3, 413–25.

Boom, B.M. (1985). 'Advocacy botany' for the Neotropics. *Garden*, 9(3), 24–32.

Boom, B. M. (1987). Ethnobotany of the Chácobo Indians, Beni, Bolivia. Advances in Economic Botany 5. The New York Botanical Garden. New York.

Boom, B. M. (1990). Useful plants of the Panare Indians of the Venezuelan Guayana. *Advances in Economic Botany*, 81, 57–76.

Boster, J. S. (1984). Classification, cultivation, and selection of Aguaruna cultivars of *Manihot esculenta* (Euphorbiaceae). Advances in Economic Botany, 11, 34–47.

Camargo, Felisberto, C. (1951). Reclamation of the Amazon floodlands near Belém. In *Proceedings of the UN Scientific Conference on*

the Conservation and Utilization of Resources, Vol. VI: Land Resources, pp. 598–602. United Nations, New York.
Campbell, D. G., Daly, D. C., Prance, G. T., and Maciel, U. N. (1986). Quantitative ecological inventory of terra firme and várzea tropical forest on the Rio Xingu, Brazilian Amazon. *Brittonia*, **38**, 369–93.
Carneiro, R. (1983). The cultivation of manioc among the Kuikuru of the upper Xingú. In *Adaptive responses of native Amazonians* (ed. W. Vickers and R. Hames).
Chernela, J.M. (1986). Os cultivares de mandioca na área do Vaupés (Tukano). pp. 151–158. In *Suma etnológica Brasileira 1*. Etnobiologia, (ed. D. Ribeiro). FINEP, Petropolis.
Cochrane, T. F. (1984). Amazonia: a computerized overview of its climate, landscape and soil resources. *Interciencia*, **9**, 298–306.
Denevan, W. M. and Padoch, C. (eds) (1987). Swidden fallow agroforestry in the Peruvian Amazon. *Advances in Economic Botany*, **5**.
Dodson, C. H. (1966). Ethology of some Euglossine bees. *Journal of the Kansas Entomological Society*, **39**, 607–29.
Dodson, C. H. (1967). Relationship between pollinators and orchid flowers. *Atas. Simp. Biota Amazônica*, **5** (Zoologia), 1–72.
Dressler, R. L. (1968*a*). Observations on orchids and Euglossini bees in Panama and Costa Rica. *Review of Biology in the Tropics*, **15**, 143–85.
Dressler, R. L. (1968*b*). Pollination by Euglossine bees. *Evolution*, **22**, 202–10.
Fearnside, P. M. (1988). Jarí at age 19: Lessons for Brazil's silvicultural plans at Carajás. *Interciencia*, **13**, 12–24.
Fearnside, P. M. and Rankin, J. M. (1980). Jarí and development in the Brazilian Amazon. *Interciencia*, **5**, 146–56.
Fearnside, P. M. and Rankin, J. M. (1982). The new Jarí: risks and prospects of a major Amazonian development. *Interciencia*, **7**, 329–39.
Fearnside, P. M. and Rankin, J. M. (1985). Jarí revisited: changes and the outlook for sustainability in Amazonia's largest silvicultural estate. *Interciencia*, **10**, 121–9.
Gentry, A. H. (1988). Tree species richness of upper Amazonian forests. *Proceedings of the US National Academy of Sciences*, **85**, 156–9.
Goodland, R. J. A. and Irwin, H. S. (1975). Amazon jungle: green hell to red desert? Elsevier, Amsterdam.
Goulding, M. (1980). *The fishes and the forest*. University of California Press, Berkeley, Los Angeles.

Goulding, M. (1989). *Amazon, the flooded forest.* BBC Books, London.

Hecht, S.B. (1989). Indigenous soil management in the Amazon Basin: some implications for development. In *Fragile lands of Latin America: strategies for sustainable development* (ed. J. O. Browder), pp. 16–81. Westview Press, Boulder, Colorado.

Herndon, W. L. and Lardner, G. 1854). *Exploration of the Valley of the Amazon.* Taylor and Maury, Washington D.C., (Reprinted 1952, New York: McGraw Hill).

Jordan, C. J. and Herrera, (1981). Tropical rain forests: are nutrients really critical? *American Naturalist,* **117,** 167–80.

Kubitzki, K. (ed.) (1983). *Dispersal and distribution: a symposium.* Sonderb. naturwiss. Ver. Hamburg 7.

Kubitzki, K. (1985). Dispersal of forest plants. pp. 192–206. In *Amazonia: key environments* (ed. G. T. Prance and T. E. Lovejoy). Pergamon Press, Oxford.

Lathwell, D. J. and Grove, T. L. (1986). Soil–plant relationships in the tropics. *Annual Review of Ecology and Systematics,* **17,** 1–16.

Milliken, W. (1991). *The ethnobotany of the Waimiri—Atroari Indians of Brasil.* Royal Botanic Gardens, Kew.

Mori, S. A. and Prance, G. T. (1990). Taxonomy, ecology and economic botany of the Brazil nut (*Bertholletia excelsa* Humb. and Bonpl.: Lecythidaceae). *Advances in Economic Botany,* **8,** 130–50.

Myers, N. (1990). *Deforestation rates in tropical forests and their climatic implications.* Friends of the Earth, London.

Nelson, B. W., Absy, M. L., Barbosa, E. M., and Prance, G. T. (1986). Observations on flower visitors to *Bertholletia excelsa* H.B.K. and *Couratari tenuicarpa* A.C. Sm. (Lecythidaceae). *Acta Amazonica,* **15,** (1/2). Supplemento 225–34.

Padoch, C., Chota Inuma, J., de Jong, W., and Unruh, J. (1985). Amazonian agroforestry: a market-oriented system in Peru. *Agroforestry Systems,* **3,** 47–58.

Pannell, C. M. (1989). The role of animals in natural regeneration and the management of equatorial rain forests for conservation and timber production. *Commonwealth Forestry Review,* **68,** 309–13.

Peters, C. M. and Hammond, E. J. (1990). Fruits from the flooded forests of Peruvian Amazonia: yield estimates for natural populations of three promising species. *Advances in Economic Botany,* **8,** 159–76.

Peters, C. M. and Vasquez, A. (1987). Estudios ecológicos de camu-camu (*Myrciaria dubia*). I. Producción de frutos en poblaciones naturales. *Acta Amazonica,* **16–17,** 161–73.

Peters, C. M., Balick, M. J., Kahn, F., and Anderson, A. B. (1989). Oligarchic forests of economic plants in Amazonia: utilization and conservation of an important tropical resource. *Conservation Biology*, **3**, 341–361.

Pires, J. M. (1966). The estuaries of the Amazon and Oyapoque Rivers. In *Proceedings of the Dacca Symposium*, pp. 211–18. UNESCO.

Posey, D. A. (1982). The Keepers of the forest. *Garden*, **6** (1), 18–24.

Posey, D. A. (1983). Indigenous ecological knowledge and development of the Amazon. In *The dilemma of Amazonian development*, (ed. E.F. Moran), pp. 225–58. Westview Press, Boulder, Colorado.

Posey, D. A. (1984). A preliminary report on diversified management of tropical forest by the Kayapó Indians of the Brazilian Amazon. *Advances in Economic Botany*, **1**, 112–26.

Prance, G. T. (1976). The pollination and androphore structure of some Amazonian Lecythidaceae. *Biotropica*, **8**, 235–241

Prance, G. T. (1985). The pollination of Amazonian plants. In *Amazonia: key environments*, (ed. G. T. Prance and T. E. Lovejoy), pp. 166–191. Pergamon Press, Oxford.

Prance, G. T. (1989). Give the multinationals a break. *New Scientist*, **1683**, 62.

Prance, G. T. (1990a). Consensus for conservation. *Nature*, **345**, 384.

Prance, G. T. (1990b). Fruits of the rainforest. *New Scientist*, **1699**, 42–5

Prance, G. T., Rodrigues, W. A., and da Silva, M. F. (1976). Inventário florestal de uma hectare de mata de terra firme km 30 Estrada Manaus–Itacoatiara. *Acta Amazonica*, **6**, 9–35.

Prance, G. T. Balée, W., Boom, B. M., and Carneiro, R. C. (1987). Quantitative ethnobotany and the case for conservation in Amazonia. *Conservation Biology*, **1**, 296–310.

Salati, E. and Vose, P. B., (1984). Amazon Basin: a system in equilibrium. *Science*, **225**, 129–38.

Salati, E., Vose, P. B., and Lovejoy, T. E. (1986). Amazon rainfall, potential effects of deforestation and plans for future research In *Tropical forests and world atmosphere* (ed. G. T. Prance), pp. 61–74. Westview Press, Boulder, Colorada.

Setzer, A. W., Pereira, M. C., Pereira, A. C., and Almeida, S. A. O. (1988). *Relatorio de atividades do Projeto IBDF-INPE 'SEQE'—Ano 1987*. Instituto de Pesquisas Espacias, São José dos Campos, São Paulo.

Shukla, J., Nobre, C., and Selers, P. (1990). Amazon deforestation and climate change *Science*, **247**, 1322–5.

Smith, N. J. H. (1981) *Man, fishes, and the Amazon*. Columbia University Press, New York.

St John. T. V. (1985). Mycorrhizae. In *Amazonia: key environments*. (ed. G. T. Prance and T. E. Lovejoy), pp. 277–83. Pergamon Press, Oxford.

Veja (1991). Uma vitória verde para os brasileiros. *Veja*, 9 January 1991, 43–4.

8
The natural world: a global casino
John Phillipson

Dr John Phillipson, Chairman of the Royal Society for Nature Conservation (RSNC), is a former Fellow and now Emeritus Fellow of Linacre College. As Reader in Animal Ecology at Oxford University he was responsible for important research in energy flow through ecosystems which broke new ground. Following retirement from his Readership in 1987, John Phillipson devoted much of his time to a study of the neglect of the natural world. He spent two years travelling throughout the world on behalf of the World Wide Fund For Nature, leading a team charged with assessing the value and progress of a wide variety of conservation projects. His final report, which was highly critical of the ill-advised use of resources by a number of governments and organizations, attracted considerable publicity.

'Quot homines, tot sententiae'—'So many men, so many opinions'
(Publius Terentius Afer, 185–159 BC)

There is general agreement that the biosphere—that part of the planet in which organisms live and reproduce—is being adversely affected by net increases in some of its atmospheric gases. A 'greenhouse effect' has been postulated and it is predicted that by the year 2050 AD the world's mean temperature will have risen by between 1.5 and 4.5°C. During the same period it is estimated that mean sea-level will increase by some 25–35cm (Barkham and MacGuire 1990). Pronounced shifts in rainfall are anticipated. The implications of such climatic changes are that there will be both modification and redistribution of the earth's biological resources.

Man is rightly concerned about the probable impact of the atmospheric changes he has unwittingly wrought; even so, he would do well to remember that desertification and flooding are not new phenomena—they can result from overgrazing and irresponsible forest clearing as well as from global warming. There can be little doubt that over many millenia man has been inexorably effecting changes on a regional scale not dissimilar to those he now fears will occur world-wide over a relatively short period of time. Current popular concern, however, focuses mainly on probable future—not past or present—changes in ecosystems. The perceived danger is that man-induced rates of change will become so rapid that the long-term biospheric adjustments necessary to achieve some form of global stability will not be possible without a major re-shuffle of the natural resources of the biosphere.

The natural resources of the biosphere are, in effect, assets; as such they can be categorized as either *fixed* or *current*. The *fixed* assets are the non-living (abiotic) components, exemplified by gases (the atmosphere), water bodies (the hydrosphere), and solid inorganic matter (the lithosphere); together these constitute the physico-chemical environment. The *current* assets are the living (biotic) components—a potentially renewable stock of plants (flora) and animals (fauna). Transfers within and between the two major types of asset can, and do, take place; for example, the daily exchanges of heat energy between atmosphere, hydrosphere, and lithosphere and also the biological processes of photosynthesis and decomposition which involve energy transformations and exchange of chemical elements between abiotic and biotic parts of the biosphere.

On a global scale the concept of the earth being a single system is not difficult to comprehend. The material resources are finite, and significant amounts of matter are neither lost nor gained across the boundary between atmosphere and space. Earth is essentially a closed system with respect to matter but an open one so far as energy is concerned (Phillipson 1975). Radiant energy from the sun enters the biosphere and is re-radiated to space as heat. The maintenance of global stability requires that the biospheric inputs and outputs of energy equal each other over time; if this equality is severely disrupted then unstable conditions will persist until the changed amounts of input and output equalize and a new equilibrium is achieved. Global warming is a clear indication of unstable, non-equilibrium conditions; the problem faced by man is not whether a new equilibrium will eventually be reached but whether, when it is, conditions will be suitable for man's existence.

On a local or regional scale every ecosystem—be it on land or in the ocean—is, like the biosphere, a functioning system. Unlike the biosphere, however, significant amounts of matter can be lost or gained across boundaries (which, it must be admitted, are frequently difficult to define). Ecosystems smaller than the biosphere are essentially open systems with respect to matter as well as energy. Left unperturbed over ecological or evolutionary time the constituent ecosystems of the biosphere will, as a result of interactions between organisms and environment, also reach a state of equilibrium; classical examples of this are mature tropical forests and well-established coral reefs. Because of the dynamic nature of the interactions between living and non-living components, ecosystems smaller than the biosphere rarely achieve a fixed and lasting equilibrium, and instead exhibit varying degrees of fluctuation (Phillipson 1989a).

The virtually closed biosphere is clearly a mosaic of many interacting smaller systems in which the sum of the parts is more stable than any one of the constituent parts. Biospheric stability and local ecosystem stabilities are inextricably linked; on these grounds alone a strong case can be made for protecting the earth's natural ability to regulate its own stability by maintaining habitat diversity.

Management of the biosphere's present habitat diversity and natural resources is multinational; 174 nations currently oversee global assets which include 1841 thousand million metric tonnes dry mass of plant material (Phillipson 1973). On a world basis this stock, if distributed evenly, would be no more than one centimetre thick and proportionately thinner than the skin of an apple. The plant assets are not distributed evenly however; the oceans of the world, for example, occupy 71 per cent of the earth's surface but, at any one time, account for only 0.2 per cent of the plant stock. In contrast, the land masses—occupying 29 per cent of the surface area—hold 99.8 per cent of the world's stock of vegetation. Superficial, and it must be stressed erroneous, interpretation of these figures would suggest that the oceans are of little import as repositories of the biosphere's current assets. Nothing could be further from the truth: the almost 500-fold difference in the quantities of stock supported at any one time is by no means a measure of the relative productivities of land and sea.

Stocks, being retained assets, do not represent profit, whereas turnover of stock with adequate return does. The turnover rates of marine and terrestrial plant stock differ markedly. Oceanic vegetation yields, on average, 14.1 times its own dry mass annually; terrestrial vegetation has a much slower turnover with a yearly yield of a mere sixteenth part or 6.2 per cent of the annual mean mass of stock. Put another way, some 70 per cent of the earth's surface, the oceans, produces 30 per cent of the world's annual plant profit. It follows that 30 per cent of the earth's surface, the land, generates 70 per cent of the total planetary plant profit per year.

On a unit area basis, average annual plant production on land is five times greater than that in the oceans. It is clear in both absolute and unit area terms that the land is more productive than the oceans; the evidence supporting the need for conservation of habitat diversity on land could hardly be stronger. However, not all of this quoted production is directly beneficial to man; it is important to recognize that overall productivity should not be confused with the yields that man can utilize.

Most of our fish protein comes from the nutrient-rich coastal and upwelling waters which occupy 10 per cent of the ocean surface, equivalent to 36 million square kilometres. The sustainable harvest of sea foods that can be taken from these areas is estimated to be 20 million tonnes dry mass annually, a level already reached on occasion by the world's fishing fleets. Animal protein from land sources largely derives from savanna, temperate grasslands, and tundra which, collectively, cover 32 million square kilometres; the estimated sustainable harvest of meat from this source amounts to 3 million tonnes dry mass only (Phillipson 1973). In absolute—and unit area—terms the protein-producing areas of the oceans are between six and seven times more productive than those on land. Protein is not, however, the whole diet; and for completeness one must add the annual 9.1 thousand million tonnes dry mass of edible crops from 14 million square kilometres of cultivated land. On this reckoning the land produces 450 times more edible produce of use to man than the oceans.

The land therefore, with its diverse flora and fauna, is increasingly recognized as a precious asset which man must manage well or perish. No longer should we plead for nature conservation on aesthetic or ethical grounds alone.

In November 1990 the Rt. Hon. Chris Patten, MP, declared: 'Conserving biodiversity is definitely in our economic interest. It is enlightened self-interest. Our survival may depend on biodiversity. We at last recognise Whitehead's "false dichotomy" "to think of nature and man".' In the same speech he also stated: 'At the end of the day how much is conserved will depend on economic and political considerations as well as on ethical or scientific decisions.' (Patten 1990). We should rejoice that politicians are, at long last, recognizing the need for concern for the environment and placing it on their agenda; but the important question remains what governments will do in practice to conserve and protect it.

In matters of the environment we can never presume that the people and governments of nations will be like-minded. Attitudes differ dramatically; at one extreme there is concern only for the interests of people like oneself; at the other, idealistic dreaming of a world that is never to be. Attitude—either of individual, institution, or government—naturally varies according to the environmental problem under review and is largely determined by the adjudged relative importance of factors such a economics, scientific knowledge, social conscience, public opinion, international pressure, and economics. Subsequent action based on what is euphemistically termed 'informed opinion' does not invariably produce adequate safeguards; frequently, the reverse is the case. Man is demonstrably gambling with the planet's natural resources and it is not unreasonable to think of the biosphere as a betting-shop or, perhaps more sophisticatedly, as a global casino.

It is not difficult for governments, whatever their persuasion, to subscribe to a general philosophy for the conservation of nature; lip-service is not expensive and the costs of attending international conventions are not prohibitively high. Conventions are generally rich in rhetoric but poor on protocols. It took some four years (1981–85) to prepare and agree the Vienna Convention for the protection of the ozone layer, and even then international agreement could not be reached on a protocol restricting the use of chlorofluorocarbons (CFCs). Five further rounds of negotiations were needed before a limited response could be formulated and agreed in the Montreal Protocol of 1987. A fully revised Protocol was finally agreed at a meeting in London during 1990. Imagine, nine years to reach agreement on the reduction of ozone-depleting substances. A mere nine months later (20 February 1991) the *Independent* newspaper displayed the headline 'Greenhouse pledge is at risk', referring of course to a pledge made by the UK concerning reductions in carbon dioxide emissions.

Undaunted by the inability to agree on relatively simple issues, politicians pursue even more complex ones. In 1989, the then Prime Minister of the UK told the UN General Assembly that the UK supported the need for a global convention on biological diversity. Mr Chris Patten informs us that the Government hopes that this will be ready for the 1992 UN Conference on Environment and Development (Patten 1990), more grist to a slowly grinding mill. There is a world of difference separating principle and practice; perhaps there is more than a little truth in a Third World view that conservation is a game played largely for the benefit of the more affluent members of the 'Western Club'.

International pressures, of course, influence the decision-making of national governments poor and rich; but certain national data-sets are available to help gauge the problems which a country that declares for conservation faces. They are:

- the area of land available;
- the population pressure on that land as reflected by population density and population growth;
- infant mortality rate as a measure of medical and social need;
- the per capita Gross National Product (GNP);
- the area of land legally protected as nature reserves.

A recent study of 22 non-European countries (Phillipson 1989b) revealed that the size of their 'conservation estates' ranged from 0.5 per cent of the total area of the country (Mexico) to 15 per cent (Tanzania). Mexico's record is adjudged as being poor because the average value for Latin America was shown to be 4.5 per cent. Conversely, with an average value for Sub-Saharan Africa of 6 per cent, the commitment to conservation shown by Tanzania is laudable.

Nobody could doubt that a multiplicity of factors influence the size of any country's 'conservation estate'; it seems reasonable to expect that countries with lower than average population pressures and a better than average per capita GNP within their continental region should have a better than average conservation estate. Not all countries meet these expectations. In China, for example, the population density is lower than the average for Asia but the size of the conservation estate is, against expectation, also lower—poor marks for this. Population growth is also lower than the average for Asia—poor marks again as the expectation would be a larger than average conservation estate. A lower than average infant mortality per 1000 live births in China suggests that demands on limited resources for medical and social care are not as great as elsewhere, leaving some leeway for the advancement of conservation by the establishment of a larger than average conservation estate—still more poor marks. By way of contrast, per capita GNP which is lower than the average for Asia leads to the expectation of a lower than average conservation estate—on this occasion expectations are met.

Using size of conservation estate and data related to four socio-economic indices, a system of scoring one for 'worse than expected', two for 'expected', and three for 'better than expected' was adopted (Fig. 8.1). China, for example, was allocated a score of five out of a possible maximum of 12. All countries were treated in a similar manner. Ecuador, Costa Rica, and Nepal were amongst the highest scorers while Brazil, China, and Madagascar vied with one another for the lowest score.

Commitment to conservation by countries can be further explored when the allocation of resources to nature conservation, namely manpower and money, are well documented. Expenditure and number of men per unit area, like the socio-economic indices, can be compared with regional averages. By

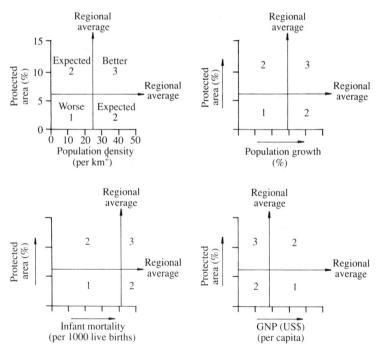

Fig. 8.1 Size of 'conservation estate' and its relationship to socio-economic indices.

employing the same technique as used in the semi-quantitative exploration of socio-economic factors, a further possible maximum of six points can be added to the scores already achieved. With the additional scoring Ecuador (Galapagos Islands), Nepal, and Senegal shot to the top while China, Peru, and Madagascar fared less well. It must be said that these results confirm one's subjective assessment of conservation care in the various countries.

Commitment to conservation, including sustainable development objectives, appears to be strongest when:

- an influential leader declares it should be so, for example Indira Ghandi in India;
- non-government agencies actively promote conservation;
- local people become involved in conservation projects;
- local people benefit either financially or in kind as a result of conservation activities;
- the country itself makes a substantial contribution in cash or kind to conservation.

Countries with little obvious commitment to conservation should not expect wealthier ones to bail them out on every occasion, especially when they voice strongly-held views about neo-colonialism and declare that they know what to do within their own countries. 'Gain not pain' is the cry, 'give us the money and we can do the job.' It is indeed unfortunate that some of them cannot. Life and living is an everlasting lottery.

Public attention is all too easily captured and held by environmental problems which are either global in nature, affect one's domestic comfort, or have furry, wide-eyed animal appeal. I have no argument with the organizations and institutions which deploy international or domestic 'flagships' in pursuing—and raising funds for—the well-being of our atmosphere, land, and seas, or the survival of the living organisms which inhabit them. I do believe however that there is a danger that local and national issues of relevance to our immediate environment are not given as much attention as they should be. It is sensible to remind ourselves from time to time that the state of the natural world in the UK, as elsewhere, is linked to wider environmental issues (RSNC 1990).

The UK has a remarkable wealth of habitats within its boundaries. Such wealth not only reflects diverse geology, soils, and topography, but also thousands of years of human occupation. Even so, it is climate that has given these islands their specialness when compared with continental Europe. An island location, a warming Gulf Stream, and moisture-laden westerly winds have provided us with a unique and diverse selection of natural habitats, ranging from blanket-bog in the far west and north to heathland and broad-leaved woodlands in the east and south (Ratcliffe 1990).

Some 5000 years ago, at the end of the so-called Atlantic Period, almost the whole of Britain was covered with virgin forest which was just beginning to be affected by human activities. Today the woodland cover of England and Wales approximates to 8 per cent of the total land surface while farmland, in all its guises, occupies some 72 per cent of the total area. The remaining 20 per cent can be accounted for by water and wetlands (1 per cent), urban areas (9 per cent), and a miscellany of roads and other man-manipulated features (10 per cent). There can be no denying that the UK is one of the least-wooded countries in Europe. Moreover, most of the woodland present today is recent, planted within the last 50–100 years, and largely at the expense of either the little ancient woodland remaining, or areas of blanket-bog such as that in the 'Flow' country of northern Scotland. There has been catastrophic loss over recent years of ancient woodland. Between 1945 and 1985 about 10 per cent was grubbed out for agriculture; a further 30–40 per cent was changed drastically by the planting of conifers and other species of less value to native wildlife.

The national catalogue of wildlife disasters is not restricted to ancient woodlands. In 1984 the Nature Conservancy Council (NCC) estimated that 80 per cent of the species-rich lowland grasslands of sheep walks on chalk and Jurassic limestone had been lost, mostly in the last 45 years (NCC 1984). An extreme case can be found in Berkshire where only 0.33 per cent of the original chalk outcrop remains as semi-natural chalk downland. Another drastic loss of habitat type can be highlighted in Oxfordshire where only 2.5 hectares of heathland remain, a mere 0.001 per cent of the county's area. It is sobering to learn also that there were country-wide annual losses of some 2900 miles of hedgerows between 1969 and 1980 and of 4000 miles between 1980 and 1985. Attrition is clearly the order of the day (RSNC 1989).

As Sir William Wilkinson, the Chairman of the now-dismembered Nature Conservancy Council said in March 1989:

'The more the wider environment becomes impoverished in its wildlife, the sharper is the distinction between what is special and what is not. And the more that natural and semi-natural habitats decline through human impact, the more important do the remaining areas become.'

Some six to seven years ago the statutory government agency for nature conservation in Britain made clear that at least 10 per cent of the area of England, Scotland, and Wales must be protected for nature conservation under law, and that this figure may require upward revision if the country-wide losses of nature continue at an unacceptable level (NCC 1984). Currently about 8 per cent of Great Britain has been designated as Sites of Special Scientific Interest (SSSIs).

The SSSI network was designed to ensure that sufficient sites in kind, number, and extent are protected, in order to conserve the total national resource of the range of variation in habitats, with their associated flora and fauna. Can we be satisfied that this is the case? Is legislation protecting our special sites or not?

Unfortunately it is abundantly clear that current legislation, 'The Wildlife and Countryside Act 1981', does not afford effective protection for SSSIs, let alone for non-designated areas of the countryside. Nowhere is safe—Lord Caithness, a Minister within the Department of Environment, wrote in a letter to RSNC dated 24 July 1989:

'As I am sure you are aware, our aim is to provide for the necessary development and economic growth without detriment to our wildlife, but in a crowded island such as ours, it is inevitable that conflicts of interest will arise. The system we have created by our Wildlife and Countryside Act 1981 and Department of Environment Circular 27/87 "Nature Conservation" ensures that each legitimate interest is given full consideration before a decision is made. I remain to be convinced that there is a better way of protecting our natural heritage.'

If this is the best that the Government can offer then the outlook is indeed bleak. Some four weeks prior to the penning of Lord Caithness' letter, Virginia Bottomley, a then Parliamentary Under-Secretary of State for the Environment, stated in response to a Parliamentary Question (in June 1989) that between 1 April 1984 and 31 March 1988, 687 SSSIs had sustained damage, some 14 per cent of the then notified sites. I can further report that between 1 April 1989 and 31 March 1990 a further 300 SSSIs were damaged, 39 of them permanently (NCC 1990).

Of great significance in the above context is the fact that the 'Wildlife and Countryside Act 1981' *excludes* any damaging operation carried out on a SSSI if it has the benefit of planning permission. It must be said that 'blanket' planning permissions for mineral and peat extraction (Interim Development Orders) —issued during the late 1940s and the 1950s—take precedence over SSSI designation (RSNC 1991). Moreover, some Local Planning Authorities, while ostensibly giving due regard to nature conservation interests, authorize development of sites (SSSIs) designated as being of national network importance.

Politicians regularly invoke the word 'balance'—a weasel word. It is lack of balance the public should deplore. Rhetoric apart, the environment still carries relatively little weight on the see-saw of politics; development is all the cry, but development at all costs is the road to destruction. Before criticizing others one should make sure that one's own house is in order.

Statutory government agencies with responsibility for environmental matters have little money and few teeth. If they are to protect the natural heritage of England, Scotland, and Wales and now perhaps Northern Ireland as well, they will need all the backing and support they can get. They depend, more than most people realize, on the voluntary conservation organizations. If you care for our natural world then support and help to influence the statutory organizations by becoming supporters of the voluntary ones. Partnership is the name of the game. Join your local 'Wildlife Trust', one of 47 partners in 'RSNC: The Wildlife Trusts Partnership'. There is no better way of 'monitoring the environment'.

REFERENCES

Barkham, J. and MacGuire, F. (1990). The heat is on. *Natural World*, **30**, 24–8.
Department of the Environment (1981). *Wildlife and Countryside Act.* Royal Assent Oct. 30, 1981.
NCC (1984). *Nature conservation in Great Britain*, pp. 111. Nature Conservancy Council, Peterborough.
NCC (1990). *Sixteenth report* pp. 184. Nature Conservancy Council, Peterborough.
Patten, C. (1990). Conserving biological diversity. *NERC News*, January 1991, 1–11.
Phillipson, J. (1973). The biological efficiency of protein production by grazing and other land-based systems. In *The biological efficiency of protein production* (ed. J.G.W. Jones), pp. 217–35. Cambridge University Press, Cambridge.
Phillipson, J. (1975). Introduction to ecological energetics. In *Ecological Bioenergetics* (ed. W. Grodzinsky, R. Z. Klekowski, and A. Duncan), pp. 3–13. IBP Handbook No.24. Blackwell Scientific Publications, Oxford.
Phillipson, J. (1989*a*). *Urban ecosystems : soils, soil fauna and productivity*. UNESCO Int. Sci. Workshop on Soils and Soil Zoology in Urban Ecosystems, pp. 103–23. Deutsches MAB-Nationalkomitee, Bonn.
Phillipson, J. (1989*b*). *Evaluation of the effectiveness of WWF projects*, pp. 203 and 51pp of appendices. World Wildlife Fund, Gland, Switzerland. (Restricted circulation).
Ratcliffe, D. (1990). Treasures of these isles. *Natural World*, **30**, 12–16.

RSNC (1989). *Losing ground : habitat destruction in the UK*, pp. 21. RSNC, Lincoln.

RSNC (1990). *Nature conservation : the health of the United Kingdom*, pp. 24. RSNC Lincoln.

RSNC (1991). *Losing ground : skeletons in the cupboard—mineral planning*, pp. 22. RSNC, Lincoln.

Wilkinson, W. (1989). Foreword in *Guidelines for selection of biological SSIs : rationale and operational approach and criteria*. Nature Conservancy Council, Peterborough.

Index

açaí palm 178
acid rain 26–8, 43, 96
acoustic remote sensing 151–2
Acre extractivist reserve 172, 173
Agassiz, E. and L. 159–60
Aghulas current 123, 125, 139, 139–40
agoutis 165
agriculture 99
  carrying capacity and mode of 13
  floodplain 179–80
  integrated 25
  intensification 25
  see also land
agroforestry 174–9
aguaje palm 181
air-conditioning 11, 37
ALARA (As Low as Reasonably Achievable) 22, 32
ALATA (As Low as Technically Achievable) 32
Albritton, D. L. 28
algae 113–14
  see also plankton
Allegretti, M. H. 172
Amapá Territory 158, 159, 173
Amazon rainforests 157–87
  alternatives to deforestation 170–82
  conservation 183–5
  ecological background 161–5
  historical background 158–61
  political disagreements 185–7
  rate of deforestation 157–8, 166
  short-term unsustainable approach 165–8
  sustainability 168–70
  value of cooling effect 116–17
America 159, 160
  see also United States
Anderson, A. B. 178
Anderson, M. 26

andiroba oil 172
anglers' lead weights 20–1
animal–plant interactions 164–5
Antarctic Circumpolar current 135–7
Antarctic ice sheet 88–9, 90
Argos system 152
Ashby, E. 26
Atlantic Isopycnic Model (AIM) 155
Atlantic Ocean 126
atmosphere
  carbon dioxide in 56–9
  modelling 67–70, 109, 128–9
  doubling carbon dioxide 70–9
  sensitivity 79–81
  TOGA model 143
  pollution 96
  predictability 140–1
  storms and ocean storms 129, 130–1
  see also acid rain; greenhouse effect; global warming; ozone layer depletion
ATTOM 151
autonomous, unmanned submersibles (autosubs) 123, 154–5, 156

babassu palm 182
Balée, W. 170
Balick, M. J. 182
Barbier, E. B. 25
Barco, President 183, 185
Barkham, J. 193
Baruch Commission 101
BATNEEC (Best Available Technology Not Entailing Excessive Costs) 32
Battarbee, R. W. 27
beef production 117
  see also cattle pasture

bees 164–5
Belgium 160
Bergen Ministerial Conference (1991) 44
BET (Best Environmental Timetable) 33
biocidal pollutants
 man-made 25
 natural 15–24
biodiversity 99, 100
 governments and conserving 197, 198
biogeochemical cycles 14
biogeochemistry 108
biological reserves 183–5, 186
biosphere 193–5
 assets 194
Birkhead, M. 21
blood lead levels 20
Bolivian Syndicate 160
Boom, B. M. 170, 171
Bora Indians 174, 175, 176, 177, 178
Boster, J. S. 178
Bottomley, V. 204
BPEO (Best Practicable Environmental Option) 32
Brazil 116, 161
 conservation policy 166
 ownership of Amazonia 158, 159, 160
 see also Amazon rainforests
Brazil nuts
 extractivist reserves 172
 pollination of trees 164–5
buriti palm 181

Cabano war 159
Caithness, Lord 203–4
Camargo, F. 179
camu-camu 181
Canadian Climate Centre model 71
cancer, fear of 109
car manufacturing 19
Carajás project 183
carbo 97
 compounds as pollutants 21

cycle 14, 56
 see also carbon dioxide; chlorofluorocarbons
carbon dioxide 55
 in the atmosphere 56–9
 effect on radiation budget 59–62
 global warming potential 29
 marine algae and 114
 model simulations for doubling 70–9
 and temperature 70–5, 78, 112
 time-dependent simulations of predicted emission scenarios 84–7, 88
 UK targets for reducing 50–1, 91
Carneiro, R. 178
carrying capacity 12–13
cattle pasture 165–6
 see also beef production
Central Electricity Generating Board (CEGB) 27–8
CERMA beer factory 182
Chácobo Indians 170
chalk downlands 202–3
Challenger 126
chemistry 108
Chernela, J. M. 178
Chernobyl disaster 24, 31
China 102
 conservation
  commitment to 200, 201
  estate 199
chlorofluorocarbons (CFCs) 28, 109–10, 119–20
 disposal 110
 global warming 29, 63–4
 Montreal Protocol 28, 35, 64, 91, 198
cholera 110–11
Clarke, R. 22, 34
Clay, J. 172
Clean Air Acts 26, 43
climate convention 100
climate models 3, 66–83
 see also atmosphere; oceans
clouds 116
 and algae 113–14
 modelling effects of 79–80
coal 112–13

Index

Cochrane, T. F. 164
co-evolution 107
Collor, President 187
Committee on Biological Effects of Ionising Radiation 22
commitment to conservation 200–1
Commonwealth Conference, Kuala Lumpur 45
communist countries 46–7
computers
  climate models and 3, 69
  ocean circulation forecasting and 147
Conference on Security and Co-operation in Europe 45
conservation
  government attitudes towards 197–9
  commitment to 200–1
  rainforests 183–5, 186
conservation estates 199
Consumers' Association 37
  see also green consumerism
Conway, G. R. 25
Cook, Captain 125
copaiba tree 172
Costa Rica 200
Council of Europe 47
CRAY YMP supercomputer 3, 69
crop diversity 178
crude oil, *see* oil
currents, ocean 125–6, 127
  measuring 151–2
  modelling 131, 132–8, 139
    heat circulation 139–40
    topography and 134–8
  see also oceans
Czechoslovakia 46–7

Denevan, W. M. 174
desertification 193–4
Dodson, C. H. 165
Dolphin Autosub 154–5, 156
Drax power station 27
Dressler, R. L. 165
drinking water 20

earth
  fragility of 105–6
  scientific views of 106–8
eastern Europe 46–7
Ecological Security Council 101
economic growth 53–4
ecosystems 195
  as view of earth 107
Ecuador 200, 201
Ehrlich, A. H. 8
Ehrlich, P. R. 6, 8
El Niño-Southern Oscillation (ENSO) 143–4, 146, 148
Elkington, J. 45
energy 98–9, 194–5
  alternative sources 99
  consumption 6, 8–10
    means of reducing 10–11
  pricing 103
  subsidy and food 11
engineering 110–11
Environment Protection Act (1990) 49
environmental law 34–5, 36
environmental organizations 48, 204–5
environmental residence times 16, 17
environmentalist business management schemes 31
environmentally friendly products 37, 45–6
  see also green consumerism
Euglossine bees 164–5
European Community 34, 47–8
European Ocean Observing Satellite (ERS-1) 149
Everest, D. A. 34
expansion, thermal sea-water 89, 90
  see also sea level rise
extractivist reserves 171–3

Fearnside, P. M. 168
Fieux, M. 128
Fine Resolution Antarctic Model (FRAM) 124, 132–40
  heat circulation 139–40

(FRAM) (*Cont.*)
  plankton ecology 140
  simulations compared with observations 139
fish
  protein 196
  rainforest 180
floats 152
flooding 193–4
floodplain agriculture 179–80
food 11–13
Fordlândia 167
forests
  destruction 99, 115
  tropical 115–17
  see also Amazon rainforests; woodland
fossil fuels 98
  global warming 29, 55
  non-polluting ways of burning 112–13
  rate of emission of carbon dioxide 56–7
  reliance on 10, 95
fragility of earth 105–6
France 159

Gaia theory 2, 107
Gardner, M. J. 24
Gentry, A. 161
Geophysical Fluid Dynamics Laboratory Model 71
geophysiology 108–9
  view of the earth 107
Germany 43
Gilfillan, S. C. 16
glaciers 87–90
Global Change Programme 107
Global Climate Observing System 98
  International Planning Office 148
Global Ocean Observing System (GOOS) 125, 148, 149, 151
global warming 29–30, 55–91, 96, 195
  effect of carbon dioxide and water vapour 59–62
  effect of other greenhouse gases 62–4
  international action 35, 96
  model simulations/predictions 66–83
  observed temperature record 64–5
  sea level rise 87–90
  time-dependent simulations and effect of oceans 83–7
  see also greenhouse effect; temperature
GLOSS network 151
gmelina trees 167–8
Goddard Institute of Space Studies model 71
goldminers 166–7
Goodland, R. J. A. 179
Goulding, M. 180
Great Britain, *see* United Kingdom
green consumerism 37, 45–6, 48, 51–2
greenhouse effect 55–91, 96, 148, 193
  atmospheric carbon dioxide 58–9
  global carbon dioxide budget 56–8
  predictability of oceans 144–5
  radiation budget
    effect of carbon dioxide and water vapour 59–62
    role of other greenhouse gases 62–4
  reductions of emissions needed 97
  see also carbon dioxide; global warming
Greenland ice sheet 88–9, 90
Green movement 30, 36, 45, 110, 111
Grove, T. L. 164
growth, economic 53–4
Gulf oil pollution 2, 141–3
Gulf Stream 139
Guruswamy, L. D. 34, 35, 37

Hadley Centre 49
Hailes, J. 45
Hammond, E. J. 181

Index

heat
  circulation in oceans 139–40
  radiation on atmosphere 59–64
  see also global warming; temperature
heathland 203
Hecht, S. B. 178
hedgerows 203
Herndon, W. L. 159
Herrera, R. 164
Holdgate, M. W. 13
Holdren, J. P. 6
Holligan, P. M. 113
Houghton, J. T. 87, 89, 96
Hudson Institute 160
human will 3–4, 121–2
Hutchinson, G. E. 108
hydrogen as fuel 113

ice ages 95
ice sheets/cover 80, 87–90, 97
India 102
Indian Ocean 139
Indians' system of tree usage 170–1
  see also Amazon rainforests
individual everyday decisions 3–4, 36–7
industrial revolution 95–6
industry
  environmental standards 52, 100
  Green movement's view of 111
infra-red radiation 61–2
  see also greenhouse effect
Integrated Pollution Control (IPC) 49
interdependence 2
Intergovernmental Panel on Climate Change (IPCC) 96, 100, 101
  greenhouse effect 87, 89, 144
international action 100–2
international agreements 34
  see also under individual names
International Court of Justice 101
International Geosphere Biosphere Programme 87, 107
International Institute for Environment and Development 7
International Tropical Timber Organization (ITTO) 171
Irwin, H. S. 179
Italy 43

Jardim, M. A. G. 178
Jári forestry project 167–8
Jasanoff, S. 35
Jivaro Indians 178
Jones, R. R. 24
Jordan, C. J. 164
jute 179

Ka'apor Indians 170
Kahn, H. 160
Kasten, R. 161
Kayapó Indians 178
King, J. L. 34
Kramer, L. 36
Kuala Lumpur Commonwealth Conference 45
Kubitzki, K. 164
Kuikuru Indians 178

La Condamine, C. M. de 158
land
  carrying capacity 13
  effect of industrial revolution 95–6
  Ministries of Land Use 99
  plant production 196
  protein production 196–7
  see also agriculture
Lardner, G. 159
Lathwell, D. J. 164
Law of the Sea (UNCLOS) 34, 35, 100
lead 16–21, 33
Lee, T. R. 24
Leggett, J. 29

leukaemia 24
Linet, M. S. 24
Londer, R. S. 107
Ludwig, D. 167, 168
Lutzenberger, J. 185, 187

MacGuire, F. 193
Madagascar 200, 201
Madrid Treaty 158
man
  impact on the environment 6–30, 193–4
    energy 8–11
    food 11–13
    pollution, *see* pollution
    population 6–8
  role as steward 121–2
man-made biocides 25
Manabe, S. 65, 84, 85, 86
marine biology 113–15
marine chronometer 123, 125
Marshall, Lord 27–8
Mason, B. J. 27, 29, 30, 56, 83
mauritia trees 181
Maury, M. F. 159
media-led public outcry 33–4
medical treatments, radioactivity and 22
Mendes, F. (Chico) 171–2
mercury 166–7
mesoscale upwelling 131–2, 132
Mestrino, G. 187
Meteorological Office climate models 66–83
  atmospheric model 67–70
  ocean circulation 82
  sensitivity 79–81
  simulations for doubling carbon dioxide 70–9
methane 63
methanol 113
Mexico 199
Milliken, W. 170
Mineraçào Rio Norte bauxite mine 183
mining 182–3
  see also goldminers

Mitchell, J. F. B. 70, 80
Mitterand, F. 161
modelling, mathematical, *see* atmosphere; climate models; oceans
Ministries of Land Use 99
Moldán, B. 46
Molina 108
monocultures, forestry 167–8
Montreal Protocol 28, 35, 64, 91, 198
Morgan, A. 24
Mori, S. A. 165
mountain ranges, submerged 134–8
Murozumi, M. 16
Myers, N. 29, 169
mycorrhizal fungi 163

National Biogeochemical Ocean Fluxes Study 87
National Coal Board (NCB) 27
National Radiological Protection Board (NRPB) 22
national sovereignty 52
natural biocides 15–24
Natural Environment Research Council (NERC) 123
  Autosub 123–4, 154–5, 156
Nature Conservancy Council (NCC) 49–50, 202, 203, 204
Nepal 200, 201
Nelson, B. W. 165
NIABY (Not In Anyone's Back Yard) 30–1
NIMBY (Not In My Back Yard) 30–1
nitrous oxide 63
non-reversible degradation 6
nuclear power 10, 31
  radioactive pollution from 22–4
nuclear waste 24

oceanographic atlases 128, 134
oceans
  algae and climate 113–14

oceans (*cont.*)
  instruments inside 151–5
    autosubs 154–5
    drifting 152
    fixed 151–2
    mobile 152–4
  mathematical modelling 129–32
    coupled to atmospheric models 69, 82
    delaying effects 83–7, 97
    FRAM 132–40
    operational forecasting of circulation 145–8
    predictability 140–5
    WOCE 145
  monitoring 123–56
    classical analysis 127–9
    last 200 years 125–7
  plant production 195–6
  pollution 96
  protein production 196–7

  tools for observing 148–55
  see also sea level rise
oil
  crude and salt marshes 15
  Gulf pollution 2, 141–3
oligarchic forests 180–3
orchids 164–5
O'Riordan, M. 22
oxygen, rainforest production of 116
ozone layer depletion 28, 45, 49, 96, 198
  see also chlorofluorocarbons

Padoch, C. 174
paint, lead in 20
Panama Canal 115
Panero, R. 160
Pannell, C. M. 164
paper pulp 167–8
Patten, C. 197, 198
Perrins, C. M. 20–1
Peru 158, 201
  see also Amazon rainforests
pesticides 25
Peters, C. M. 180, 181

petrol, lead in 17–20, 33
Phillipson, J. 194, 195, 199
planetary medicine 112
plankton 131–2, 132, 140
planning permission, SSSIs and 204
plant–animal interactions 164–5
plant production 195–6
politics of the environment 3–4, 43–54
'polluter pays' principle 28, 33
pollution 13–30, 95–6
  disturbance of physical–chemical systems 25–30
  Gulf oil 2, 141–3
  man-made biocides 25
  natural biocides 15–24
  see also acid rain; global warming; ozone layer depletion
Pollution Inspectorate 49
population growth 6–8, 95
Porter, G. 28
Portugal 158
Posey, D. A. 178
poverty 7–8
power stations, pollution from 26–8, 48, 112
  see also nuclear power
Prance, G. T. 164, 165, 170, 183, 185
precipitation
  changed patterns 30, 97
  deforestation in Amazonia 169
  modelling doubling of carbon dioxide 71, 76–7, 78–9
  see also acid rain
precision hydrography 153–4, 155, 156
Proudman Oceanographic Laboratory 151, 154
public concern 45
  attitudes 30–1
  media-led outcry 33–4
  politicians' task 53–4
  presssure for change 52–3
  see also green consumerism

radiation, heat, *see* heat
radioactivity 15, 21–4

radiometers, scanning 149
radon 221
rainfall, *see* precipitation
'Rainforest Crunch' 172
rainforests 29, 115–17
　Amazon, *see* Amazon rainforests
　cooling effect 116–17
Rankin, J. M. 168
Ratcliffe, D. 202
refugees 102–3
Renberg, I. 27
Rennell, J. 123, 125–6
Rennell Centre for Ocean
　　Circulation 123, 126
reserves, rainforest 183–5, 186
Reynolds equation 129
rice 179
river-flow reversal 34
Roberts, L. E. J. 10, 28
Roosevelt, F. D. 160
Rossby radius of deformation 129
Rowlands 108
Royal College of Radiologists
　　(RCR) 22
Royal Commission on
　　Environmental Pollution
　　(RCEP) 9, 15, 24, 34
BET 33
　BPEO 32
　lead pollution 16, 17–18, 19
　NIMBY 30, 31
　pesticides 25
Royal Society 32
RSNC 202, 203, 204
rubber cultivation 167, 171–2
rubber tappers 171–2

Salati, E. 169, 183
Sansavini, S. 25
satellites 149–51
Scandinavia 26, 27–8
scanning radiometers 149
Schneider, S. H. 107
scientific views of earth 106–9
sea level rise 30
　estimates 87–90, 97, 193
sea level gauges 151

Sea Soar 155
Sears, J. 21
Seasat mission 149
Second World Climate
　　Conference 98, 100, 101
Sellafield reprocessing plant 33–4
Senegal 201
Setzer, A. W. 166
Seveso chemical plant 43
ships, research 152–4
Shukla, J. 169
Sites of Special Scientific Interest
　　(SSSIs) 203–4
Smith, N. J. H. 180
smog 43
snow cover 80, 97
soil 162–4
solar energy 99
Southwood, T. R. E. 6, 12, 19, 30, 32
　acid rain 27
　pesticides 25
　radioactivity 22, 24, 34
sovereignty, national 52
'spaceship earth' concept 107–8
species diversity, *see* tree species
　　diversity
species interactions 164–5
St John, T.V. 163
statutory conservation agencies
　　204–5
Stefan's Law 61
stewards of the earth 121–2
Stommel, H. 128
storms 30
　ocean 138
　　atmospheric and 129, 130–1
straw-burning 33
Surface Water Acidification
　　Programme (SWAP) 27, 33
sustainable development 44
　Amazon rainforests 168–70
Swallow, J. 127, 152
Swallow float 123, 152
swans 20–1

Tamshiyacu 174–8, 179
Tanzania 199

Index 215

targets, realistic 50–1
Tarnavskii, A. 34
temperature
  carbon dioxide and 112
    modelling effects of
      doubling 70–8
    time-dependent simulation of
      emission scenarios 84–7, 88
  expected increase 96–7, 193
  measured global 64–5
  measuring by satellite 149
  see also global warming
Thatcher, M. H. 94–5, 198
thermal expansion of sea-water 89, 90
This common inheritance
    (Government White
    Paper) 48, 50, 94, 103
Thompson, W. 126
thresholds, pollutants and 14–15
timber extraction 171
TOGA model 143, 152
Tolba, M. K. 28
topography, ocean and
  currents 134–8
toxic wastes 43
transport 99, 103, 113
tree species diversity 161–2
  uses of 170–1
  see also Amazon rainforests
tropical rainforests, *see* Amazon
  rainforests;
  rainforests
Tukano Indians 178
typhoid 110–11

umari trees 174
Union of Soviet Socialist Republics
    (USSR) 34
United Kingdom 2, 48, 198
  acid rain 26–8
  Amazon rainforests 159, 160, 161
  carbon dioxide emissions 50–1, 91
  commitment to conservation
    201–5
  environmental legislation 43,
    49–50, 203–4

  green consumerism 46, 48
  lead in petrol 17–18, 19–20
  public concern 45–6
  radiation dose variations 23
  see also This common inheritance
United Nations 94
United Nations Conference on the
    Environment and
    Development 44
United Nations Convention of the
    Law of the Sea
    (UNCLOS) 34, 35, 100
United Nations Environment
    Programme 7, 101
United Nations General
    Assembly 101
United Nations Security Council 101
United States 9, 161
  environmental legislation 34–5
  National Academy of Sciences 96,
    98
  pollution 97, 101–2
  see also America

várzea forest 179–80
Vasquez, A. 181
Vernadsky, V. I. 107
Vienna Convention for the
    protection of the ozone
    layer 198
Vivaldi Operation 155
voluntary conservation
    organization 48, 205
Vorsorgeprinzip principle 32, 33
Vose, P. B. 169

waste disposal
  CFCSs 110
  nuclear 24
  toxic 43
water
  lead in drinking 20
  pollution 96
  see also oceans; sea level rise
Watson, A. 113, 114

Webb, D. J. 134
Whitfield, M. 113
Wildlife and Countryside Act
    (1981) 203–4
Wilkinson, W. 203
Wilson, C. A. 70
Winter, G. 31
woodland 202
Workshop 90, 185, 186
World Bank 31
World Climate Research Programme
    (WCRP) 56, 82, 101, 144,
    145
World Commission on the
    Environment and
    Development 44
World Conference on the
    Environment 100
World Meteorological
    Organization 101
World Ocean Circulation
    Experiment (WOCE) 56, 83,
    123, 145
    computers 147
    hydrographic stations 145, 146
    measuring instruments 152, 153
World Resources Institute 7
World Weather Watch (WWW) 141,
    148
Wynne, B. 24

Yanomami Indians 161, 166–7
Yanomono 161